Luciano André de Aguiar Pedrosa
Elisabeth C. Silva
José Gustavo Melo

Land use dynamics in the municipality of Tamandaré, Pernambuco, Brazil

Luciano André de Aguiar Pedrosa
Elisabeth C. Silva
José Gustavo Melo

Land use dynamics in the municipality of Tamandaré, Pernambuco, Brazil

Case study on a rural property

ScienciaScripts

This book is a translation from the original published under ISBN 978-613-9-64866-5.

Publisher:
Sciencia Scripts
is a trademark of
Dodo Books Indian Ocean Ltd. and OmniScriptum S.R.L publishing group

120 High Road, East Finchley, London, N2 9ED, United Kingdom
Str. Armeneasca 28/1, office 1, Chisinau MD-2012, Republic of Moldova, Europe
Printed at: see last page
ISBN: 978-620-7-79901-5

ACKNOWLEDGEMENTS

I thank God and our Lord Jesus Christ for always being present in our lives. To my wife (Ana Rosa) for her understanding, patience and encouragement in completing this work. To my mother who has always supported me at every moment of my life. In memory of my father who is in another constellation spiritually guiding me. To my course colleague and advisor for the completion of the CBT, Profa. M. a. Elisabeth Regina Alves Cavalcanti and to my co-supervisor Prof. M. e. José Gustavo da Silva Melo.

SUMMARY

This work brings together technical elements and analyses of the Geoprocessing and Georeferencing set, and preparation of the CAR for a property located in Engenho Barreirinho in the municipality of Tamandaré - PE, owned by UMBUZEIRO EMPREENDIMENTOS IMOBILIÁRIOS LTDA - CNPJ: 10.876.584/0001-17 250,500 ha. - Latitude: 08°44'06.78" S Longitude: 35°19'31.5" O. In this sense, the aim of this work is to analyse the dynamics of land use in the municipality of Tamandaré/PE using geoprocessing and remote sensing techniques on a rural property in the area corresponding to Engenho Barreirinho - Rural Zone of the municipality of Tamandaré/PE, demonstrating the effectiveness of the conventional topographic method using Geodetic GPS together with Nautical GPS in places of closed forest. This was made possible with the help of Remote Sensing and Geoprocessing in the preparation of the CAR for a property located in Engenho Barreirinho in the municipality of Tamandaré, located on the south coast of Pernambuco. A vegetation index (SAVI) based on a Landsat 5 image was used to characterise the land use profile in the municipality of Tamandaré, showing that most of the municipality has a moderate to high degree of anthropisation. Once the property had been geo-referenced on site, the CAR was drawn up containing the specific features of the property and a descriptive memorial and plan were drawn up.

Keywords: Georeferencing; Vegetation Index; Anthropisation; Descriptive Memorial.

SUMMARY

CHAPTER 1

INTRODUCTION

The development of agriculture and the use of land in a way that not only protects it from surface alterations caused by the constant action of natural phenomena, but also gradually develops its productive capacity, requires careful initial planning (SILVA, 2009). By analysing characteristics such as vegetation cover, topography, drainage and soil type, it is possible to arrive at the rational and appropriate use of a given geographical space (SILVA, 2011a; SILVA, 2016). In this way, areas for the preservation of water sources, forest reserves, agricultural areas, industrial districts and urban expansion areas are determined, so that land use obeys the natural characteristics of the landscape, and planning considers sustainable development (TUCCI, 1993).

According to DIEGUES (1989), environmental planning is based on the need to incorporate the environmental variable into socio-economic planning, with a view to making the most appropriate use of ecosystem space and its resources. Thus, planning begins with the collection of basic information, which includes: general environmental characteristics, soil and water resources, natural forests, wildlife, fishing resources, soil erosion and the condition of the various environmental units, land and land use patterns, human cultures and communities, pollution, pests and diseases, the impacts of human activities on the natural and human environment (SILVA, 2015).

According to ROCHA (1977), there is a need for surveys for land use planning purposes. It is the integration of these surveys that will provide the conditions for the development of the rural environment. For FUCHS (1986), the term land use can be understood as the way in which space is being occupied by man. A land use survey consists of mapping and assessing qualitatively and quantitatively everything that exists on the lithosphere.

In this respect, according to COELHO (1971), a land use survey is a study aimed at assessing soil resources in terms of their capacity, location and the estimation of suitable or poorly utilised land. ROSA (1994) states that the data provided by land use surveys helps to forecast crops and their commercialisation, assess forest cover and its changes, or determine new areas for agricultural and forestry expansion.

In this sense, the use of geotechnologies has made important contributions to monitoring forest cover and land use in recent decades Novo (1989). The environmental damage caused by uncontrolled anthropogenic action affects ecosystems and populations. However, the use of technologies that facilitate a broader diagnosis more quickly, such as remote sensing and especially geographic information systems (GIS), can help to solve various problems linked to the interests and needs of the communities involved (CONCEIÇÃO, 2004). Thus, for planning to be effective and efficient, it is necessary to have access to correct and detailed information, which can be available in aerial remote sensing images (SILVA, 2011a).

The development of remote sensing techniques has made it possible to acquire a variety of information about the earth's surface, supporting temporal, edaphic and phenological analyses of vegetation (VINAGÓ et al, 2011). Pereira et al (1989) state that a survey of current land use, which is necessary for planning purposes, can be obtained using multispectral data provided by remote sensing satellites, combined with interpretation techniques.

Man's occupation of space is associated with the appropriation of natural resources, which in recent decades has led to increased interest in environmental issues, such as the preservation of the environment, as well as its maintenance and recovery. In this context, it is necessary to identify and understand the forms of space in order to be aware of land use and occupation in a given area and in different periods (FACCO et al., 2016).

According to Valente and Castro (1983), one of the difficulties in planning is reconciling conservation programmes with economic exploitation. Often, the owners of the latifundia or minifundia that occupy certain areas, with rare exceptions, are not very sensitive to aspects of soil, vegetation and water conservation. In this sense, the application of the vegetation index technique has facilitated various studies on monitoring land use, with the aim of highlighting variations in vegetation density, facilitating the identification and mapping of vegetated and non-vegetated areas (MENESES, 2011). Using remote sensing, it is possible to analyse, among other variables, the chlorophyll content of vegetation using the normalised difference vegetation index (NDVI) (SILVA, 2015b).

In order to perceive changes in a given environment, it is necessary to use satellite images from different dates, thus obtaining a multi-temporal analysis. Studies on multitemporal analysis using satellite images are becoming increasingly popular, serving to monitor urban growth as well as the evolution of deforestation and agricultural extension, among others (CARVALHO JÚNIOR et al., 2005).

To corroborate work based on satellite images, topographic surveying is an important tool. Knowledge of an area by means of a graphic representation defining the size, contour, relief, natural features, details such as buildings and the relative position of a part of the earth's surface are frequent concerns of professionals responsible for urban and rural planning and projects, or all those who need to know the elements that characterise an area. When this graphic representation refers to a restricted part of the earth's surface, the problem is the subject of Topography (SILVA, 2009).

We can say that the main objective of topography is to collect data (take measurements of angles, distances and unevenness) so that a portion of the earth's surface can be represented on an appropriate scale. The operations carried out in the field, with the aim of collecting data for subsequent representation, are called topographic surveys (VEIGA et al., 2007). Thus, the purpose of topography is to determine the contour, size and relative position of a limited portion of the earth's surface, without taking into account the curvature resulting from the earth's sphericity (ESPARTEL, 1987).

In this sense, the aim of this work is to analyse the dynamics of land use in the municipality of Tamandaré-PE using geoprocessing and remote sensing techniques on a rural property in the area corresponding to Engenho Barreirinho - Rural Zone of the municipality of Tamandaré/PE, demonstrating the effectiveness of the conventional topographic method using Geodetic GPS together with Nautical GPS in places of closed forest. This will be possible with the help of Remote Sensing and Geoprocessing in the preparation of the CAR for a property located in Engenho Barreirinho in the municipality of Tamandaré, located on the south coast of Pernambuco.

CHAPTER 2

OBJECTIVES

2.1 General

To analyse the dynamics of land use in the municipality of Tamandaré-PE through geoprocessing and remote sensing on a rural property corresponding to Engenho Barreirinho - Rural Zone of the municipality of Tamandaré/PE.

2.2 Specific

> Demonstrate the effectiveness of the conventional topographic method, using GEODETIC GPS in conjunction with the NAUTIC in closed forest sites, with the aid of Remote Sensing and Geoprocessing;

> To carry out a spatio-temporal analysis of the municipality of Tamandaré-PE, using a vegetation index;

> Draw up the Rural Environmental Registry (CAR) for the rural property of Engenho Barreirinho;

> Draw up a descriptive memorial for the rural property of Engenho Barreirinho;

> Draw up a plan of the rural property of Engenho Barreirinho;

CHAPTER 3

THEORETICAL BACKGROUND

3.1 Meaning of Topography

One of the main objectives of topography is to determine the relative coordinates of points. To do this, they need to be expressed in a coordinate system. There are basically two types of system used to unequivocally define the three-dimensional position of points: Cartesian coordinate systems and spherical coordinate systems (FAGGION, 2012).

According to NBR 13133 (ABNT, 1991, p. 3), the Brazilian Standard for Surveying, surveying is defined as:

"A set of methods and processes which, by measuring horizontal and vertical angles, horizontal, vertical and inclined distances, with instruments suitable for the desired accuracy, primarily establish and materialise support points on the ground, determining their topographic coordinates. These points are linked to detail points with a view to their exact planimetric representation on a predetermined scale and their altimetric representation by means of contour lines, with a predetermined equidistance and/or dimensioned points."

According to Garcia and Piedade (1979), Topography is based on the study of the description of a limited part of the earth's surface, developing hypotheses to graphically represent its horizontal projection. Doubek (1989) mentions that Topography aims to study the instruments and methods used to obtain a graphic representation of a portion of the land on a flat surface.

According to Cuiabano (2006), the aim of Topography is to study the methods used to obtain a graphic representation, on an appropriate scale, of the results of field survey activities (taking measurements of angles, distances and unevenness), to determine the contours, dimensions and relative positions in certain areas of land, in such a way as to make it possible to represent a portion of the earth's surface on a flat surface, without taking into account the curvature resulting from the earth's sphericity.

According to BRINKER and WOLF (1977), practical surveying work can be divided into five stages:

1) Decision-making, where survey methods, equipment, positions or points to be surveyed, etc. are listed;

2) Field work or data acquisition: taking measurements and recording data;

3) Calculations or processing: making calculations based on the measurements obtained to determine coordinates, volumes, etc;

4) Mapping or representation: producing the map or chart from the measured and calculated data.

Generally speaking, a total station is nothing more than an electronic theodolite (angular measurement), an electronic distance meter (linear measurement) and a mathematical processor, combined into a single unit. Based on information measured in the field, such as angles and distances, a total station can, according to FAGGION (2012), obtain other information such as:

D Reduced distance to the horizon (horizontal distance);

J Unevenness between points (point "a" equipment, point "b" reflector);

C Coordinates of the points occupied by the reflector, based on a previous orientation.

In addition to these facilities, this equipment allows corrections to be made when measurements are taken or even pre-programmed for the automatic application of certain parameters such as:

- ❖ Environmental conditions (temperature and atmospheric pressure);
- ❖ Prism constant. It is also possible to configure the instrument according to the needs of the survey, changing values such as:
- ❖ Instrument height;
- ❖ Reflector height;
- ❖ Unit of angular measurement;
- ❖ Unit of measurement of distance (metres, feet);
- ❖ Origin of the vertical angle measurement (zenithal, horizontal, nadir, etc).

3.2 Clarifications on Georeferencing

Although today spatial localisation is a relatively simple task for the satellite positioning user, it was one of the first scientific problems that human beings tried to solve. Conquering new frontiers and travelling safely required mastery of the art of navigation, knowing how to get from one place to another, with knowledge of your position throughout the journey, whether on land, at sea or in the air (ROQUE et al. 2006).

In recent years there has been great progress in the development of techniques and technologies in the areas of surveying, topography, geodesy, cartography, geoprocessing, remote sensing, aerophotogrammetry, etc. Over time, constant technical and methodological developments have led to improvements in the planning and execution of topographic and geodetic services. Developments and updates to the equipment used in topographic and geodetic surveys have improved measurement methods, be they conventional, with theodolites, levels, total stations, GNSS receivers, etc (VOGEL et al., 2011).

The first classifications of land use were based on field work. According to Oliveira (2005), in colonial Brazil, land ownership could only be obtained in two ways: legally by donation from the Portuguese crown and illegally by invading properties. Over the years, the right to property began to be acquired legally through the purchase and possession of vacant and unproductive areas, and illegally through land grabbing. Subsequently, since the 1950s, a large number of researchers in various parts of the world have dedicated themselves to the detailed identification of agricultural crops in aerial photographs (STEINER, 1970; BORGES, 1993).

With the aim of curbing fraudulent land grabbing and the illegal creation of large estates, Law 10267 was created on 28 August 2001 to establish the Public Land Registry System. Since the law came into force for the cadastre and registration of real estate, certification of ownership will only be fully achieved if the descriptive memorial of the land boundaries is signed by a qualified professional with the appropriate ART Technical Responsibility Annotation, containing the coordinates of the vertices defining the boundaries of rural properties, georeferenced to the Brazilian Geodetic

System and with positional accuracy set by INCRA at 50 cm or better (PARZZANINI, 2007).

The geo-referenced description guarantees that there will be no overlapping of properties, as long as the established accuracies are met, avoiding the creation of overlapping property titles with different owners, as in the past. The requirement for a Technical Responsibility Certificate (ART) makes professionals accountable, since by signing an ART the professional can be held liable in court for any faults in the technical procedures. It is important to note that the geo-referenced identification of the rural property area will only be required in cases of subdivision, remodelling, parceling and transfer of ownership once the deadlines set by Law 10267 have passed. The deadlines were as follows:

Ninety days for properties with an area of 5,000 hectares or more;

U One year, for properties with an area of 1,000 to less than 5,000 hectares.

These deadlines, defined by Decree No. 4.449 of 30 October 2002, came into force on the date of its publication.

- ❖ Five years for properties with an area of between five hundred and less than one thousand hectares;

- ❖ Eight years for properties with an area of less than 500 hectares.

These deadlines, defined by Decree No. 5.570 of 31 October 2005, came into force on the date of its publication, with the deadline starting on 20 November 2003.

- ❖ For properties subject to legal action, all deadlines have already passed. Once the deadlines set for the application of the law have passed, the land registry officer will only be able to carry out registry acts such as dismemberment, remapping, parceling and transfer of the total area if the technical work identifying the rural property has been certified by INCRA.

All property boundaries must be geo-referenced to the Brazilian Geodetic System and all work must be carried out using the IBGE geodetic landmarks as a reference. Usually the properties to be geo-referenced are far from the IBGE geodetic landmarks, in which case it is necessary to install a support landmark on the property that will serve as a reference for geo-referencing all the other vertices on the property.

The INCRA manual calls this landmark the Basic Support Landmark. In practice, the coordinates of the support landmark are determined using at least two IBGE landmarks and GPS signal receivers, and at the end of this determination the landmark should have an accuracy of better than 10 cm (PARZZANINI, 2007).

Another important aspect refers to the georeferencing of properties that share borders with properties already certified by INCRA. In this case, in order to avoid overlapping areas, the technical standard stipulates that the georeferenced and already certified vertices must prevail over new surveys. In this case, the professional is obliged to georeference all the vertices on the property and compare the coordinates obtained in his survey with the coordinates already certified by INCRA:

> *S* If the error found is less than 50 cm, the professional should abandon its determination and adopt the coordinates of the common points already certified by INCRA, in all calculations of area, distances and azimuth, as well as in the writing of the descriptive memorial.

> *C* If the error found shows a discrepancy greater than the permitted value, the work will not be certified by INCRA and must be reassessed by the professional in order to correct the errors in their determinations or prove any errors in the coordinates already certified.

The other vertices will be assessed based on compliance with the other procedures described in the technical standards, which must be proven by the executor through technical reports (INCRA, 2013).

In this context, GPS comes into play, which is currently the most efficient instrument for collecting spatialised point, linear and polygonal information. This is the task known as georeferencing. Georeferencing is an improved technique for describing rural properties, which helps to control both the registration of rural properties and the rights in rem relating to them. The aim of georeferencing rural properties is to specifically locate an individual property on the globe. Georeferencing the eucalyptus trees in a reforestation means obtaining the geographical coordinates of each tree; georeferencing a watercourse means travelling along it and collecting the entire route; georeferencing an area means delineating its boundaries to form a polygon. These are typical tasks for using GPS (ROQUE et al. 2006).

Given the significant amount of information that will be generated by georeferencing in rural areas, it is essential to use a geographic information system that can not only store the information generated, but also organise the cadastral database and manage spatial data in such a way as to allow system users to make the most of this information in current and future work. With this system, it will be possible to retrieve information through database queries that can be used in new work or to check the work carried out, trying to avoid duplicating efforts already made, making georeferencing work more agile and less costly (PARZZANINI, 2007).

3.3 Understanding Geoprocessing

Geoprocessing comprises various tools (BRANDALIZE, 2008), technologies and/or sciences that will be part of this study. These are: Topography, Satellite Positioning - GPS, Automated Cartography, Photointerpretation, Remote Sensing and Digital Image Processing.

According to Daianese (2001), geoprocessing transfers information from the real world to a computer system. This transfer is made on cartographic bases, using an appropriate reference system. A geoprocessing system is generally designed to process geographically referenced data, from its acquisition to the generation of outputs in the form of maps, reports or digital files (NETTO, 2001).

The coordinates of the vertices that will define rural properties must be georeferenced to the Brazilian Geodetic System (SGB), with positional accuracy set by INCRA. The coordinates must be obtained in the field from topographical surveys, GNSS (Global Navigation Satellite System) or by means of images or maps, on paper or digital, as long as they follow the SGB and INCRA (CASTRO; SILVA, 2014).

Geoprocessing and remote sensing techniques have become indispensable for obtaining information from the earth's surface, and today constitute a set of tools applicable to planning and zoning. The reliability and speed of the sensing process makes it easier to acquire data that is of great importance for mapping land use and occupation in a given region (PAULA, 2012).

3.4 Definition of Remote Sensing

Remote Sensing is defined in different ways by various authors, the most usual definition being that adopted by Avery and Berlin (1992) and Meneses (2001): "A technique for obtaining information about objects through data collected by instruments that are not in physical contact with the objects under investigation". For Jensen (2007), Remote Sensing (RS) can be defined as the art and science of obtaining information about objects without direct physical contact with the object.

Remote Sensing can also be defined as the joint use of modern sensors, data transmission equipment, data processing equipment, aircraft and spacecraft with the aim of studying the interactions in the earth's environment, without direct physical contact with the features, between electromagnetic radiation and the substances that make up planet Earth in its various manifestations (NOVO, 1995).

The potential of remote sensing images for mapping crops can be exploited both for small areas (IPPOLITI- RAMILO et al., 2003) and for extensive agricultural regions (RIZZI & RUDORFF, 2005; INPE, 2009). When properties are delimited by natural and inaccessible boundaries, following the descriptions and characteristics described in (INCRA, 2013), the chances of errors occurring in their delimitation by topographical or geodetic processes are great, which can lead to the risk of certification being cancelled, in addition to often a significant loss in the area of the property. [a] With this problem in mind, INCRA launched the possibility of delimiting natural and inaccessible boundaries through the use of remote sensing images, which are obtained from orbital or airborne sensors, thus indirectly obtaining geometric information with duly assessed accuracy and reliability.

3.5 Acceptance of the Rural Environmental Registry - CAR

CAR is an electronic register, compulsory for all rural properties, whose purpose is to integrate environmental information on the status of Permanent Preservation Areas (APP), Legal Reserve areas, forests and remnants of native vegetation, Restricted Use areas and consolidated areas of rural properties and possessions in the country. Created by Law 12.651/2012 under the National Environmental Information

System - SINIMA, the CAR is a strategic database for controlling, monitoring and combating deforestation of forests and other forms of native vegetation in Brazil, as well as for environmental and economic planning of rural properties.

According to Decree 7.830/2012 in Section II, art. 5: The CAR must include the data of the owner, rural possessor or person directly responsible for the rural property, the respective georeferenced plan of the perimeter of the property, areas of social interest and areas of public utility, with information on the location of the remaining native vegetation, Permanent Preservation Areas, Restricted Use Areas, consolidated areas and the location of Legal Reserves:

- ❖ Section II, Art. 6: Registration in the CAR is compulsory for all rural properties and possessions, is declaratory and permanent in nature, and will contain information about the rural property.

The main benefits of the CAR according to the Law are:

- ❖ Possibility of regularising APP and/or Legal Reserve natural vegetation suppressed or altered up to 22/07/2008 on the rural property, without being charged with an administrative offence or environmental crime;

- ❖ Suspension of sanctions for administrative infractions for irregular suppression of vegetation in APP, Legal Reserve and restricted use areas, committed up to 22/07/2008.

- ❖ Obtaining agricultural credit, in all its forms, with lower interest rates, as well as higher limits and terms than those practised on the market;

- ❖ Contracting agricultural insurance under better conditions than those practised on the market;

- ❖ Deduction of Permanent Preservation Areas, Legal Reserves and Restricted Use Areas from the calculation base for Rural Property Tax (ITR), generating tax credits;

- ❖ Lines of financing to support initiatives for the voluntary preservation of native vegetation, the protection of endangered species of native flora, sustainable forestry and agroforestry management carried out on rural property or possession, or the recovery of degraded areas; and

❖ Tax exemption for the main inputs and equipment, such as: wire, treated wood posts, water pumps, augers for drilling into the ground, among others used for the recovery and maintenance of Permanent Preservation Areas, Legal Reserves and restricted use areas.

CHAPTER 4

MATERIALS AND METHODS

4.1 Characterisation of the area

The study area is located in the municipality of Tamandaré, on the south coast of the state of Pernambuco, in the north-eastern region of Brazil. It belongs to the Mesoregion of Mata Pernambucana and the Microregion of Mata Meridional Pernambucana, and is located 109 kilometres south of the capital of Pernambuco (IBGE, 2013). It occupies a territorial area of 214,307 km^2 , of which 1,416 km^2 are urban perimeters. According to the IBGE (2010), its population is around 22,323 inhabitants, making it the 95th most populous municipality in Pernambuco.

4.1.1 Characterisation of the municipality of Tamandaré

Tamandaré has a hot and humid tropical climate of type As' according to the *Koeppen* climate classification. Seasonality in the region is marked by two distinct periods: a rainy period from March to August and a dry period from September to February. The highest average annual rainfall occurs in July and the lowest in December. The Tamandaré inlet is semicircular in shape, with a concavity towards the east and its main boundaries are Pontal do Lira to the north, Ponta de Mamucabinha to the south, reef formations to the east and the beach line to the west (LIMA, 1997). Three major rivers influence coastal dynamics in the region studied: the Formoso River, which is located to the north of the bay, and the Mamucaba and Una rivers to the south (ARAÚJO and COSTA, 2003). The work was therefore carried out in the locality of Engenho Barreirinho - Rural Zone of the municipality of Tamandaré/PE, Figure 1.

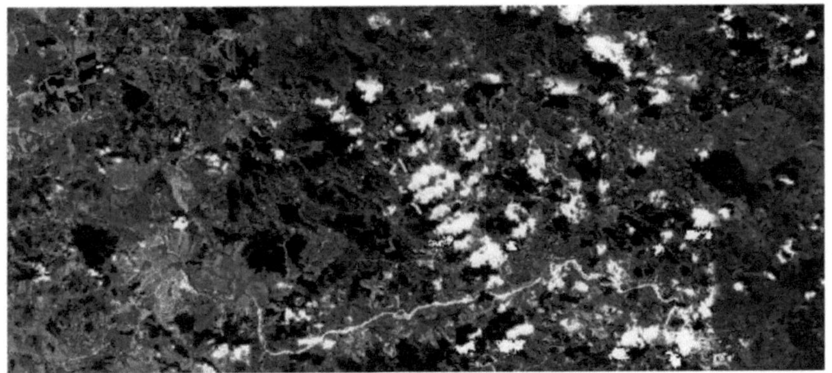

Figure 1 - CAR *Shapfile* image of the location under study: Engenho Barreirinho.
Source: Prepared by the author. CAR registration: PE-2614857-
5996.32F2.5116.444F.9DAF.70B6.E495.94AB Registration date: 27/06/2016 23:51:13

4.1.2 Characterisation of the case study area - Engenho Barreirinho

The Barreirinho Mill covers around 500 hectares and is owned by the Santo André Mill in Tamandaré, in the southern Mata region. This study will cover a property located at Engenho Barreirinho in the municipality of Tamandaré -PE, owned by UMBUZEIRO EMPREENDIMENTOS IMOBILIÁRIOS LTDA - CNPJ: 10.876.584/0001-17 250.500 ha. - Latitude: 08°44'06.78" S Longitude: 35°19'31.5" O.

4.2 Survey Description

There was a need to analyse and carry out a primary reconnaissance and a prior dialogue with the entire team involved, thus defining the division of tasks to be carried out to complete the work. The location of the vertex (V-01), where the work began, was defined, as well as the adjustments needed to carry it out, while defining the other points for closing the perimeter. The survey method applied was the classic clockwise walk with point irradiation, to define the boundaries of the Rural Property. After the perimeter was determined using a GTR A and B GPS (for the base) and a Garmin 62S GPS, for use in the closed forests and internal points of the property to determine the CAR, together with the Google Earth image and geoprocessing.

Through this information obtained in the field, we can cross-check the data for later analysis in the CAR. Carrying out this fieldwork has provided me with technical and practical knowledge that is relevant to my academic training.

The work was carried out in the locality of Engenho Barreirinho - Rural Zone of the municipality of Tamandaré/PE, where all the preparatory study for carrying out the work was carried out, involving a topography assistant, plus 4 rural workers from Engenho barreirinho, The greatest difficulty was marking the points of a particular extension of the property, as it is an area made up of closed forest to obstruct the path to delineate the perimeter studied, and certain materials were used, such as a 5kg sledgehammer, steel pointer and massaranduba wood pickets, sickle, hoe and machete, and antiophidic serum, as it is an area of venomous animals such as rattlesnakes.

The topographical survey was carried out using the anti-clockwise walking method, recording equidistant points to be determined in proportion to the bearing determined by the property manager, on vertices characterised by wooden pickets, with forced centring, which defined the topographical polygon, from whose GPS points were collected, as many as possible for the demonstration and the feature of interest to the project, as follows.

The GTR B equipment was installed as a base and fixed antenna, waiting 1 hour for a better measurement and description of the accuracy of the points after confirming the first static point at 20 minutes at point V-01, in order to continue drawing up the perimeter at vertex V-01, and then, once at vertex V-02, we proceeded to collect the points necessary for a good topographic representation of the boundaries of the area under study, recording them in a field notebook (CROQUI), and so on, until the last pre-established study vertex was occupied.

The work of processing the data collected and stored in the GPS GTR B will be carried out below. From the activities carried out in the field study, I feel the need to take care during measurement processes, especially when it comes to data collection, observing the importance of data collection, and ensuring that systematic errors and accumulative errors are minimised during field survey practices.

This work brings together technical elements and analyses of the Geoprocessing and Georeferencing set, and the preparation of the CAR for a property located in Engenho Barreirinho in the municipality of Tamandaré -PE, owned by UMBUZEIRO EMPREENDIMENTOS IMOBILIÁRIOS LTDA - CNPJ: 10.876.584/0001-17

250.500 ha. - Latitude: 08°44'06.78" S Longitude: 35°19'31.5" O.

Property details: Engenho Barreirinho - parts 1

Owner - Umbuzeiro Empreendimentos Imobiliários Ltda.

Company address: Av. Engenheiro Domingos Ferreira 467, 13° andar Pina. CNPJ: 10.876.584/0001-17

Address of the sugar mill: Estrada que liga companhia açucareira Santo André do Rio Una a Cucaú, s/n zona rural. Postcode: 55578-000.

4.3 GPS operation

The basic operating principle of GPS is to obtain the distance between 2 points (receiver and satellite), one of which has its position known and is used as a reference. The determination of an object on the earth's surface follows the principle of triangulation, where with a minimum of 3 references its position is obtained, such as its coordinates. A fourth reference adds the altitude component, allowing for greater precision in identifying and localising the object (GUERREIRO, 2002).

Satellites send radio frequency signals based on a fundamental frequency (f0) of 10.23 MHz. From this, 2 new operating frequencies are obtained by multiplying f0 by the constants 154 and 120, generating the carrier waves belonging to the L band, namely: L1 and L2 respectively. These carriers have the following characteristics, as shown in Table 1:

Table 1 - Satellite information.

Carrier	Multiplier	Operating frequency (MHz)	Wavelength
L1	154	1575,42	19.04 cm
L2	120	1227,60	24.44 cm

Source: Prepared by the author (2018).

These two carrier waves are modulated in phases, generating codes called PRN - Pseudo Randon Noise - falsely random noise, which are unique and used to identify the satellites captured.

The codes that make up the PRN are basically the C/A and P codes. The C/A (Coarse Acquisition) code is generated by a pseudo-random algorithm, using the time given by the satellites' atomic clocks. It has a frequency of 1.023 MHz (f0/10) with a wavelength of around 300 metres. It is the main component of the Standard Positioning Service (SPS) available for civilian use. It is this code that all small receivers, known as "navigation receivers", use for autonomous positioning.

The P code, which stands for Precise or Project, is transmitted at the same frequency as the fundamental frequency: $f0 = 10.23$ MHz, generating a wavelength of 30 metres. The higher frequency and shorter wavelength make this code much more precise than C/A, which is why it is reserved for military use and authorised users. It is known as Precision Positioning Service - PPS. Its generation follows complex mathematical algorithms, so that its binary frequency is repeated every 266.4 days, arranged to produce 37 sequences of unique codes, lasting 7 days, providing the existence of 37 PRNs that will identify each of the transmitting satellites.

As a security measure, it is treated with a technique known as AS- Anti-Spoofing (anti-fraud) and is also encrypted, thus becoming the Y code.

A third modulation is performed on the L1 and L2 carrier waves, giving rise to navigation margins, which are sequences of data transmitted at a rate of 5 bps (bits per second) and lasting 30 seconds, forming data frames. Each frame in turn is subdivided into 5 sub-frames or parts of 6 seconds each, containing messages as shown in Table 1.

Table 1 - Carrier wave modulation of navigation margins.

Sub-frame	Message
1	- Coefficient or parameters for satellite transmitter correction; - GPS week number; - Health of the transmitting satellite; - Age of data.
2	- Orbital parameters - broadcast or transmitted ephemerides describing the satellite's predicted and calculated position, its Kleperian orbital elements and their corrections. It consists of 16 parameters.
3	- Same function as the subframe or part 2.
4	- Information from the almanacs for satellites 25 to 32; - Models of the ionosphere, making it possible to correct the delay caused in wave transmission. - models for converting GPS time into coordinated universal time (GPST-

	UTC);
	- Information on the anti-fraud system (AS - *AntiSpoof Flag)* and configuration of the 32 satellites;
	- Satellite health 25 to 32;
	- Reserve for special messages.
5	- Information from the almanacs for satellites 1 to 24;
	- Operational conditions for satellites 1 to 24;
	- Information for adjusting satellite times.

Source: Prepared by the author (2018).

A pressing need for GPS to work properly is for the receiver and satellite to be intervisible, i.e. for there to be no or minimal obstacles between them. Radio frequency signals transmitted by satellites are able to pass through clouds, soot, dust and other less dense materials whose porosity allows the wave to penetrate. However, high-density obstacles such as forest canopies become insurmountable obstacles to the signals, preventing their reception and consequently making positioning impossible.

In this case, on the property of Engenho Barreirinho, which has a dense forest, we had to mark the last static vertex for 20 minutes and switch it off, continuing with the 62 SC Nautical GPS, accurate to 3m, starting at vertex V-0034 to V-0040, which at each vertex marking was averaged to better capture the coordinates, at the end of the forest walk, another static point was made with the Geodesic GPS at 20 minutes, also continuing to the other point of dense forest at vertex V-0045, marking another static point with the same minutes, continuing with the 62 SC nautical GPS until the end of the forest, marking another static point with the same minutes, continuing the walk until the end of the perimeter.

The radio signals transmitted by satellites have to pass through all the layers of the Earth's atmosphere until they reach the GPS receiver. During this journey, the signals are subjected to various influences that alter their characteristics, as the Earth's atmosphere has a dynamic behaviour that varies throughout the day. The troposphere causes variations that vary according to the density of the gaseous mass that is concentrated in this layer, which is accentuated in signals coming from lower elevation angles.

4.4 Satellite Image Processing

4.4.1 Space-Time Analysis

The study area was first delimited. A 2011 image from the Landsat 5 satellite was interpreted and analysed, downloaded free of charge from the website of the National Institute for Space Research (INPE) and processed at UFPE's Remote Sensing and Geoprocessing Laboratory (SERGEO), which holds the licence for the software used in this work.

4.4.2 Image processing and layout assembly

Firstly, all the images were recorded from points collected in the field. The images were downloaded from the INPE website. To process the Landsat-5 satellite images, models were created using the ModelMaker tool in the ERDAS Imagine 9.3 software.

on licence from the Department of Geographical Sciences at the Federal University of Pernambuco.

4.4.3 Radiometric calibration

The set of radiance or radiometric calibration is obtained using the equation proposed by Markham and Baker (1987) (Equation 1):

$$L\,\lambda i = \alpha t + \frac{bt - \alpha t}{255}ND \quad (1)$$

Where a and b are the minimum and maximum spectral radiances (1 1 2pm srWm), ND is the pixel intensity (integer between 0 and 255) and i corresponds to the bands (1, 2, ... and 7) of the Landsat 5 and 7 satellite. The calibration coefficients used for TM images are those proposed by Chander and Markham (2003) and Oliveira et al, 2010.

4.4. 4Reflectance

The reflectance of each band (i) is defined as the ratio between the solar radiation flux reflected by the surface and the incident global solar radiation flux, obtained using the equation (ALLEN et al., 2002 apud OLIVEIRA et al, 2010), (Equation 2):

$$\rho\lambda i = \frac{\pi . L\lambda i}{K\lambda i .\cos Z .dr} \qquad (2)$$

Where λiL is the spectral radiance of each band, Àik is the spectral solar irradiance of each band at the top of the atmosphere 12pm, Z is the solar zenith angle and rd is the square of the ratio between the mean Earth-Sun distance (ro) and the Earth-Sun distance (r) on a given day of the year (DSA), (OLIVEIRA et al, 2010; SILVA et al, 2011b).

4.4.5 *Soil Adjusted* Vegetation *Index* (SAVI)

SAVI (Equation 3) was developed by HUETE (1988) as a transformation technique to minimise the influence of soil reflectance on spectral vegetation indices involving the red and near-infrared wavelengths and to more accurately model near-infrared radiance in the more open canopies (SILVA et al, 2011b), equation 3:

$$SAVI = \frac{(1+L)(\rho IV - \rho V)}{(L + \rho IV + \rho V} \qquad (3)$$

CHAPTER 5

RESULTS AND DISCUSSION

In order to analyse the rural property located in Engenho Barreirinho, the area was first mapped using the vegetation index (SAVI) using a georeferenced Landsat 5 satellite image from 2011 on a scale of 1:40,000 (Figure 2):

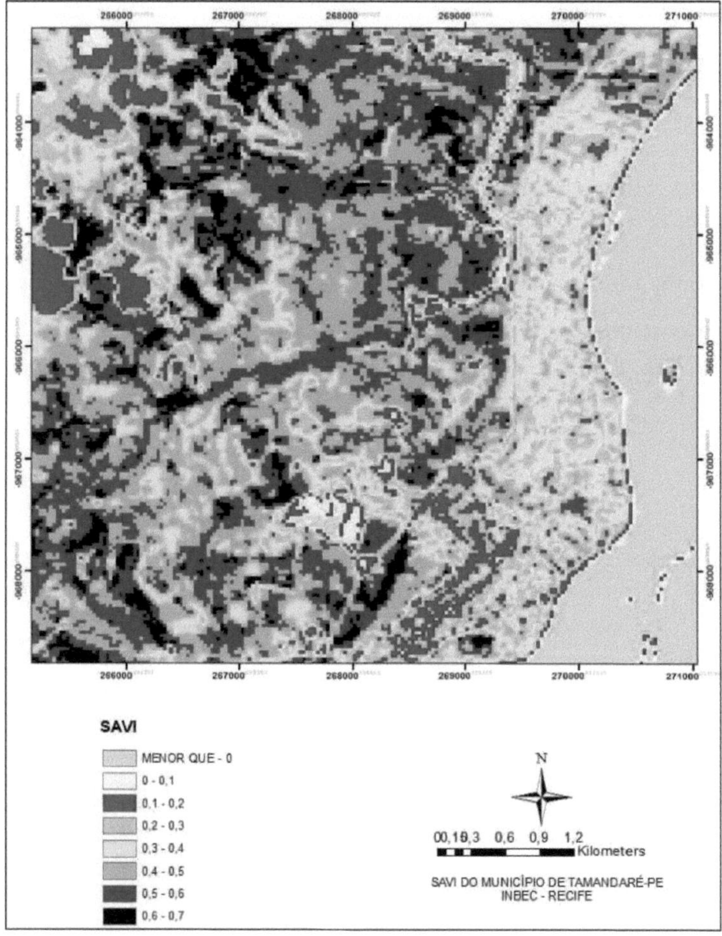

Figure 2 - SAVI of the Tamandaré-PE region.
Source: Prepared by the author (2018).

As you can see, the map above shows 8 classes, each with a different SAVI value. The SAVI results show that SAVI between 0.2 and 0.5 predominates in the area, corresponding precisely to the most anthropised areas (urban and degraded areas).

Whereas SAVI greater than 0.5 represents areas with vegetation that is being recomposed or conserved. These values are explained by the strong presence in the region of rural properties that practise agriculture (monoculture or polyculture).

Within this context is the rural property covered by this study. The image was obtained from Google Earth and the area of the property was delimited (Figure 3):

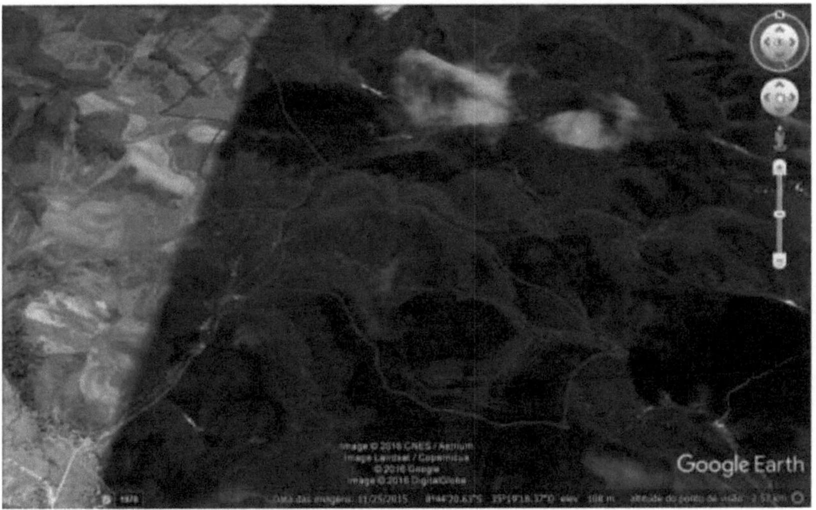

Figure 3 - Delimitation of the rural property of Engenho Barreirinho.

Source: Prepared by the author (2018).

The figure above shows the delimited area in the Google Earth image. The points needed to delimit the polygon were surveyed on site and are compiled in Table 2:

PROPERTY: ENGENHO BARREIRINHO

OWNER: GILSON LEITE CHAFAS FERREIRA

MUNICIPALITY: TAMANDARÉ

Datum: SIRGAS2000, FUSO 25 L, Central Meridian: 33 WGr

Table 2 - Analytical calculation of area, azimuths, sides, geodesic and UTM coordinates.

IMÓVEL : ENGENHO BARRERINHO
PROPRIETARIO : UMBUZEIRO EMPREENDIMENTOS IMOBILIÁRIOS LTDA
MUNICÍPIO : TAMANDARÉ
Datum: SIRGAS 2000, FUSO 25 L, Meridiano Central: 33 WGr

Estação	Vante	Coord.Norte (m)	Coord.Este (m)	Azimute	Distância Reduzida	Fator Escala	Latitude	Longitude
V-V-0001	V-V-0002	9.035.650,004	243.308,061	152°45'24"	37,163	1 00041539	8°43'00,970" S	35°19'58,742" W
V-V-0002	V-V-0003	9.035.616,964	243.325,073	152°45'24"	155,575	1 00041494	8°43'02,048" S	35°19'58,193" W
V-V-0003	V-V-0004	9.035.478,646	243.396,291	152°45'24"	569,853	1.00041328	8°43'06,563" S	35°19'55,892" W
V-V-0004	V-V-0005	9.034.972,007	243.657,154	163°27'16"	193,103	1 00041293	8°43'23,099" S	35°19'47,464" W
V-V-0005	V-V-0006	9.034.786,900	243.712,146	146°53'51"	122,487	1.00041251	8°43'29,132" S	35°19'45,703" W
V-V-0006	V-V-0007	9.034.684,310	243.779,030	152°29'43"	202,340	1.00041192	8°43'32,484" S	35°19'43,536" W
V-V-0007	V-V-0008	9.034.504,840	243.872,475	151°30'05"	229,630	1 00041122	8°43'38,342" S	35°19'40,517" W
V-V-0008	V-V-0009	9.034.303,035	243.982,040	151°30'05"	108,289	1.00041090	8°43'44,930" S	35°19'36,974" W
V-V-0009	V-V-0010	9.034.207,868	244.033,706	155°42'34"	61,117	1 00041074	8°43'48,036" S	35°19'35,304" W
V-V-0010	V-V-0011	9.034.152,161	244.058,850	145°37'39"	50,467	1 00041056	8°43'49,854" S	35°19'34,493" W
V-V-0011	V-V-0012	9.034.110,506	244.087,342	145°37'33"	6,010	1.00041053	8°43'51,215" S	35°19'33,570" W
V-V-0012	V-V-0013	9.034.105,545	244.090,735	145°32'42"	9,621	1 00041050	8°43'51,377" S	35°19'33,460" W
V-V-0013	V-V-0014	9.034.097,612	244.096,179	139°38'43"	277,845	1.00040936	8°43'51,636" S	35°19'33,283" W
V-V-0014	V-V-0015	9.033.885,881	244.276,088	137°16'52"	680,379	1.00040844	8°43'58,561" S	35°19'27,442" W
V-V-0015	V-V-0016	9.033.386,013	244.737,658	135°37'16"	260,614	1 00040529	8°44'14,917" S	35°19'12,449" W
V-V-0016	V-V-0017	9.033.199,744	244.919,932	131°44'01"	83,321	1.00040490	8°44'21,014" S	35°19'08,525" W
V-V-0017	V-V-0018	9.033.144,280	244.982,110	151°33'03"	57,521	1 00040472	8°44'22,831" S	35°19'04,503" W
V-V-0018	V-V-0019	9.033.093,706	245.009,511	132°16'30"	162,864	1 00040396	8°44'24,482" S	35°19'03,617" W
V-V-0019	V-V-0020	9.032.984,149	245.130,018	130°14'41"	235,367	1.00040283	8°44'28,071" S	35°18'59,698" W
V-V-0020	V-V-0021	9.032.832,089	245.309,673	133°45'40"	309,686	1 00040142	8°44'33,054" S	35°18'53,854" W
V-V-0021	V-V-0022	9.032.617,895	245.533,337	240°16'27"	12,825	1.00040149	8°44'40,068" S	35°18'46,582" W
V-V-0022	V-V-0023	9.032.611,535	245.522,200	170°34'36"	7,678	1.00040148	8°44'40,272" S	35°18'46,948" W
V-V-0023	V-V-0024	9.032.603,961	245.523,457	296°13'08"	7,936	1.00040153	8°44'40,519" S	35°18'46,908" W
V-V-0024	V-V-0025	9.032.607,467	245.516,338	236°34'32"	97,923	1.00040204	8°44'40,404" S	35°18'47,140" W
V-V-0025	V-V-0026	9.032.553,528	245.434,610	235°49'40"	76,767	1 00040244	8°44'42,142" S	35°18'49,824" W
V-V-0026	V-V-0027	9.032.510,409	245.371,097	242°11'48"	46,585	1 00040270	8°44'43,532" S	35°18'51,909" W
V-V-0027	V-V-0028	9.032.488,680	245.329,890	224°48'40"	159,789	1 00040341	8°44'44,231" S	35°18'53,261" W
V-V-0028	V-V-0029	9.032.375,320	245.217,275	195°18'04"	52,397	1 00040350	8°44'47,897" S	35°18'56,967" W
V-V-0029	V-V-0030	9.032.324,780	245.203,447	201°39'38"	36,486	1.00040358	8°44'49,539" S	35°18'57,430" W
V-V-0030	V-V-0031	9.032.290,870	245.189,980	195°54'45"	170,649	1.00040388	8°44'50,639" S	35°18'57,877" W
V-V-0031	V-V-0032	9.032.128,760	245.143,194	150°16'49"	25,850	1.00040380	8°44'55,969" S	35°18'59,440" W
V-V-0032	V-V-0033	9.032.104,310	245.156,009	291°21'30"	58,591	1.00040414	8°44'56,702" S	35°18'59,025" W
V-V-0033	V-V-0034	9.032.125,649	245.101,442	261°48'26"	84,421	1 00040466	8°44'55,997" S	35°19'00,806" W
V-V-0034	V-V-0035	9.032.142,923	245.018,807	290°20'59"	27,210	1.00040482	8°44'55,418" S	35°19'03,504" W
V-V-0035	V-V-0036	9.032.152,386	244.993,294	277°32'08"	10,800	1.00040489	8°44'55,105" S	35°19'04,337" W
V-V-0036	V-V-0037	9.032.153,802	244.982,588	271°24'40"	37,256	1.00040513	8°44'55,057" S	35°19'04,687" W
V-V-0037	V-V-0038	9.032.154,720	244.945,342	287°01'08"	24,893	1.00040528	8°44'55,020" S	35°19'05,905" W
V-V-0038	V-V-0039	9.032.162,005	244.921,540	297°08'36"	12,828	1.00040535	8°44'54,778" S	35°19'06,682" W
V-V-0039	V-V-0040	9.032.167,858	244.910,124	281°14'33"	13,554	1.00040543	8°44'54,585" S	35°19'07,054" W
V-V-0040	V-V-0041	9.032.170,500	244.896,830	288°31'56"	15,032	1.00040552	8°44'54,497" S	35°19'07,488" W
V-V-0041	V-V-0042	9.032.175,278	244.882,577	300°52'24"	11,611	1.00040559	8°44'54,338" S	35°19'07,953" W
V-V-0042	V-V-0043	9.032.181,296	244.872,611	289°48'03"	45,418	1 00040586	8°44'54,143" S	35°19'08,278" W
V-V-0043	V-V-0044	9.032.198,622	244.829,879	273°36'08"	45,642	1.00040614	8°44'53,633" S	35°19'09,672" W
V-V-0044	V-V-0045	9.032.199,490	244.784,327	283°40'57"	14,629	1.00040623	8°44'53,531" S	35°19'11,161" W
V-V-0045	V-V-0046	9.032.202,950	244.770,113	270°58'55"	95,720	1.00040684	8°44'53,416" S	35°19'11,626" W
V-V-0046	V-V-0047	9.032.204,590	244.674,407	246°08'51"	15,165	1.00040693	8°44'53,343" S	35°19'14,755" W
V-V-0047	V-V-0048	9.032.198,458	244.660,538	253°54'57"	6,250	1.00040696	8°44'53,342" S	35°19'15,210" W
V-V-0048	V-V-0049	9.032.196,727	244.654,533	320°00'47"	9,512	1.00040700	8°44'53,595" S	35°19'15,407" W
V-V-0049	V-V-0050	9.032.204,015	244.648,420	309°31'11"	15,371	1.00040708	8°44'53,357" S	35°19'15,605" W
V-V-0050	V-V-0051	9.032.213,796	244.636,563	295°55'39"	18,781	1.00040719	8°44'53,036" S	35°19'15,991" W
V-V-0051	V-V-0052	9.032.222,007	244.619,672	313°07'15"	39,114	1.00040737	8°44'52,765" S	35°19'16,542" W
V-V-0052	V-V-0053	9.032.248,743	244.591,122	294°22'29"	9,036	1.00040742	8°44'51,890" S	35°19'17,470" W
V-V-0053	V-V-0054	9.032.252,472	244.582,892	317°53'56"	10,176	1.00040746	8°44'51,767" S	35°19'17,738" W
V-V-0054	V-V-0055	9.032.260,023	244.576,070	312°03'50"	34,510	1.00040762	8°44'51,520" S	35°19'17,960" W
V-V-0055	V-V-0056	9.032.283,143	244.550,449	279°36'31"	36,404	1.00040785	8°44'50,762" S	35°19'18,793" W
V-V-0056	V-V-0057	9.032.289,219	244.514,556	298°23'42"	14,980	1.00040793	8°44'50,558" S	35°19'19,966" W
V-V-0057	V-V-0058	9.032.296,343	244.501,378	311°41'39"	23,112	1.00040804	8°44'50,323" S	35°19'20,395" W
V-V-0058	V-V-0059	9.032.311,716	244.484,121	317°44'26"	6,250	1.00040807	8°44'49,819" S	35°19'20,957" W
V-V-0059	V-V-0060	9.032.316,341	244.479,918	336°12'53"	13,030	1.00040810	8°44'49,668" S	35°19'21,093" W
V-V-0060	V-V-0061	9.032.328,265	244.474,662	360°00'00"	12,579	1.00040810	8°44'49,279" S	35°19'21,263" W
V-V-0061	V-V-0062	9.032.340,844	244.474,662	331°19'38"	29,459	1.00040819	8°44'48,870" S	35°19'21,260" W
V-V-0062	V-V-0063	9.032.366,690	244.460,528	339°58'11"	9,345	1.00040821	8°44'48,026" S	35°19'21,717"
V-V-0063	V-V-0064	9.032.375,470	244.457,327	324°31'44"	21,793	1.00040829	8°44'47,740" S	35°19'21,820" W
V-V-0064	V-V-0065	9.032.393,218	244.444,681	340°57'04"	10,999	1.00040831	8°44'47,160" S	35°19'22,230" W
V-V-0065	V-V-0066	9.032.403,615	244.441,091	353°30'33"	9,914	1.00040832	8°44'46,821" S	35°19'22,345" W
V-V-0066	V-V-0067	9.032.413,466	244.439,970	8°00'26"	7,968	1.00040831	8°44'46,500" S	35°19'22,380" W
V-V-0067	V-V-0068	9.032.421,357	244.441,080	356°03'53"	13,652	1.00040832	8°44'46,244" S	35°19'22,342" W
V-V-0068	V-V-0069	9.032.434,977	244.440,143	347°22'39"	11,337	1.00040834	8°44'45,800" S	35°19'22,370" W
V-V-0069	V-V-0070	9.032.446,040	244.437,666	334°26'47"	9,170	1.00040836	8°44'45,440" S	35°19'22,449" W
V-V-0070	V-V-0071	9.032.454,313	244.433,710	8°18'11"	11,495	1.00040835	8°44'45,170" S	35°19'22,576" W
V-V-0071	V-V-0072	9.032.465,688	244.435,370	335°05'37"	2,217	1.00040836	8°44'44,800" S	35°19'22,520" W
V-V-0072	V-V-0073	9.032.467,699	244.434,436	360°00'00"	2,755	1.00040836	8°44'44,734" S	35°19'22,550" W
V-V-0073	V-V-0074	9.032.470,454	244.434,436	8°24'34"	16,231	1.00040834	8°44'44,645" S	35°19'22,549" W
V-V-0074	V-V-0075	9.032.486,510	244.436,810	16°13'50"	8,515	1.00040833	8°44'44,123" S	35°19'22,469" W
V-V-0075	V-V-0076	9.032.494,686	244.439,190	348°29'13"	8,066	1.00040834	8°44'43,857" S	35°19'22,389" W

V-V-0076	V-V-0077	9.032.502.590	244.437.580	342°04'24"	13,905	1.00040836	8°44'43,600" S	35°19'22,440" W
V-V-0077	V-V-0078	9.032.515.820	244.433.300	327°04'33"	9,475	1.00040840	8°44'43,169" S	35°19'22,577" W
V-V-0078	V-V-0079	9.032.523.773	244.428.150	345°27'28"	11,474	1.00040842	8°44'42,909" S	35°19'22,744" W
V-V-0079	V-V-0080	9.032.534.880	244.425.269	333°02'32"	25,387	1.00040849	8°44'42,547" S	35°19'22,836" W
V-V-0080	V-V-0081	9.032.557.509	244.413.760	304°11'29"	28,652	1.00040864	8°44'41,808" S	35°19'23,208" W
V-V-0081	V-V-0082	9.032.573.610	244.390.060	353°47'00"	35,834	1.00040866	8°44'41,280" S	35°19'23,980" W
V-V-0082	V-V-0083	9.032.609.233	244.386.180	274°26'05"	7,239	1.00040871	8°44'40,120" S	35°19'24,100" W
V-V-0083	V-V-0084	9.032.609.793	244.378.962	323°37'54"	9,540	1.00040874	8°44'40,100" S	35°19'24,336" W
V-V-0084	V-V-0085	9.032.617.475	244.373.305	304°32'25"	6,412	1.00040878	8°44'39,849" S	35°19'24,519" W
V-V-0085	V-V-0086	9.032.621.110	244.368.024	322°40'43"	24,533	1.00040887	8°44'39,730" S	35°19'24,691" W
V-V-0086	V-V-0087	9.032.640.620	244.353.150	317°05'25"	19,292	1.00040895	8°44'39,092" S	35°19'25,173" W
V-V-0087	V-V-0088	9.032.654.750	244.340.015	354°59'48"	18,852	1.00040897	8°44'38,630" S	35°19'25,600" W
V-V-0088	V-V-0089	9.032.673.540	244.338.370	318°55'02"	20,619	1.00040905	8°44'38,018" S	35°19'25,650" W
V-V-0089	V-V-0090	9.032.689.082	244.324.820	328°21'09"	14,740	1.00040910	8°44'37,510" S	35°19'26,090" W
V-V-0090	V-V-0091	9.032.701.630	244.317.086	6°28'57"	21,820	1.00040908	8°44'37,100" S	35°19'26,341" W
V-V-0091	V-V-0092	9.032.723.310	244.319.550	340°50'27"	15,647	1.00040912	8°44'36,395" S	35°19'26,256" W
V-V-0092	V-V-0093	9.032.738.090	244.314.415	330°25'56"	24,409	1.00040919	8°44'35,913" S	35°19'26,421" W
V-V-0093	V-V-0094	9.032.759.320	244.302.370	335°06'31"	13,845	1.00040923	8°44'35,220" S	35°19'26,810" W
V-V-0094	V-V-0095	9.032.771.880	244.296.542	342°29'05"	10,181	1.00040925	8°44'34,810" S	35°19'26,998" W
V-V-0095	V-V-0096	9.032.781.589	244.293.478	335°55'15"	9,750	1.00040927	8°44'34,493" S	35°19'27,096" W
V-V-0096	V-V-0097	9.032.790.491	244.289.500	317°13'33"	9,699	1.00040932	8°44'34,203" S	35°19'27,225" W
V-V-0097	V-V-0098	9.032.797.610	244.282.914	338°33'03"	22,552	1.00040937	8°44'33,970" S	35°19'27,439" W
V-V-0098	V-V-0099	9.032.818.599	244.274.667	314°39'20"	9,302	1.00040941	8°44'33,286" S	35°19'27,704" W
V-V-0099	V-V-0100	9.032.825.137	244.268.050	335°59'54"	9,263	1.00040943	8°44'33,071" S	35°19'27,919" W
V-V-0100	V-V-0101	9.032.833.590	244.264.286	350°59'55"	14,207	1.00040945	8°44'32,796" S	35°19'28,041" W
V-V-0101	V-V-0102	9.032.847.522	244.252.064	329°47'34"	37,409	1.00040957	8°44'32,339" S	35°19'28,110" W
V-V-0102	V-V-0103	9.032.879.952	244.243.242	315°00'05"	33,517	1.00040971	8°44'31,283" S	35°19'28,719" W
V-V-0103	V-V-0104	9.032.904.052	244.219.960	301°36'21"	40,860	1.00040994	8°44'30,494" S	35°19'29,476" W
V-V-0104	V-V-0105	9.032.925.476	244.185.161	301°31'48"	11,002	1.00040999	8°44'29,790" S	35°19'30,610" W
V-V-0105	V-V-0106	9.032.931.229	244.175.783	325°31'03"	8,724	1.00041003	8°44'29,601" S	35°19'30,915" W
V-V-0106	V-V-0107	9.032.938.420	244.170.844	303°26'06"	132,634	1.00041073	8°44'29,366" S	35°19'31,075" W
V-V-0107	V-V-0108	9.033.011.500	244.060.160	295°27'20"	47,253	1.00041100	8°44'26,966" S	35°19'34,680" W
V-V-0108	V-V-0109	9.033.031.810	244.017.494	266°01'59"	16,489	1.00041110	8°44'26,297" S	35°19'36,071" W
V-V-0109	V-V-0110	9.033.030.669	244.001.044	309°52'25"	112,467	1.00041165	8°44'26,331" S	35°19'36,610" W
V-V-0110	V-V-0111	9.033.102.772	243.914.730	298°19'03"	63,367	1.00041200	8°44'23,967" S	35°19'39,418" W
V-V-0111	V-V-0112	9.033.132.830	243.858.946	37°47'47"	59,531	1.00041177	8°44'22,978" S	35°19'41,236" W
V-V-0112	V-V-0113	9.033.179.871	243.895.430	22°28'47"	27,252	1.00041171	8°44'21,455" S	35°19'40,033" W
V-V-0113	V-V-0114	9.033.205.052	243.905.850	341°23'44"	21,252	1.00041175	8°44'20,638" S	35°19'39,687" W
V-V-0114	V-V-0115	9.033.225.193	243.899.070	349°38'38"	17,356	1.00041177	8°44'19,981" S	35°19'39,905" W
V-V-0115	V-V-0116	9.033.242.266	243.895.950	358°42'38"	21,777	1.00041177	8°44'19,425" S	35°19'40,004" W
V-V-0116	V-V-0117	9.033.264.038	243.895.460	5°39'44"	13,257	1.00041176	8°44'18,717" S	35°19'40,015" W
V-V-0117	V-V-0118	9.033.277.230	243.896.768	17°27'53"	11,306	1.00041174	8°44'18,286" S	35°19'39,970" W
V-V-0118	V-V-0119	9.033.288.015	243.900.161	13°16'24"	40,892	1.00041168	8°44'17,938" S	35°19'39,857" W
V-V-0119	V-V-0120	9.033.327.815	243.909.550	6°40'20"	31,972	1.00041166	8°44'16,844" S	35°19'39,542" W
V-V-0120	V-V-0121	9.033.359.570	243.913.265	338°46'39"	87,150	1.00041186	8°44'15,612" S	35°19'39,414" W
V-V-0121	V-V-0122	9.033.440.610	243.881.717	348°09'42"	27,414	1.00041190	8°44'12,963" S	35°19'40,429" W
V-V-0122	V-V-0123	9.033.467.429	243.875.160	358°25'52"	72,452	1.00041193	8°44'12,095" S	35°19'40,638" W
V-V-0123	V-V-0124	9.033.539.740	243.870.650	351°24'21"	66,049	1.00041199	8°44'09,742" S	35°19'40,771" W
V-V-0124	V-V-0125	9.033.605.048	243.860.780	358°26'04"	10,615	1.00041199	8°44'07,615" S	35°19'41,081" W
V-V-0125	V-V-0126	9.033.615.659	243.860.490	0°17'33"	17,627	1.00041199	8°44'07,269" S	35°19'41,088" W
V-V-0126	V-V-0127	9.033.633.286	243.860.580	13°53'47"	61,790	1.00041190	8°44'06,696" S	35°19'41,081" W
V-V-0127	V-V-0128	9.033.693.268	243.875.420	15°08'13"	32,153	1.00041185	8°44'04,747" S	35°19'40,584" W
V-V-0128	V-V-0129	9.033.724.306	243.883.816	247°46'42"	13,558	1.00041192	8°44'03,739" S	35°19'40,303" W
V-V-0129	V-V-0130	9.033.719.178	243.871.265	260°37'24"	10,650	1.00041199	8°44'03,904" S	35°19'40,715" W
V-V-0130	V-V-0131	9.033.717.443	243.880.757	285°22'15"	77,102	1.00041246	8°44'03,958" S	35°19'41,059" W
V-V-0131	V-V-0132	9.033.737.880	243.786.413	278°07'23"	60,267	1.00041284	8°44'03,278" S	35°19'43,486" W
V-V-0132	V-V-0133	9.033.746.396	243.728.751	279°06'33"	56,118	1.00041319	8°44'02,989" S	35°19'45,435" W
V-V-0133	V-V-0134	9.033.755.280	243.671.340	252°14'05"	55,547	1.00041353	8°44'02,689" S	35°19'47,245" W
V-V-0134	V-V-0135	9.033.738.332	243.618.442	249°59'07"	11,838	1.00041360	8°44'03,230" S	35°19'48,979" W
V-V-0135	V-V-0136	9.033.734.280	243.607.319	258°08'06"	63,569	1.00041399	8°44'03,359" S	35°19'49,343" W
V-V-0136	V-V-0137	9.033.721.210	243.545.109	254°47'25"	29,388	1.00041417	8°44'03,772" S	35°19'51,380" W
V-V-0137	V-V-0138	9.033.713.500	243.516.750	276°23'52"	28,478	1.00041435	8°44'04,017" S	35°19'52,309" W
V-V-0138	V-V-0139	9.033.716.673	243.488.450	356°37'17"	27,585	1.00041436	8°44'03,908" S	35°19'53,234" W
V-V-0139	V-V-0140	9.033.744.250	243.487.786	331°36'47"	16,312	1.00041441	8°44'03,011" S	35°19'53,250" W
V-V-0140	V-V-0141	9.033.760.365	243.479.090	5°24'14"	13,804	1.00041441	8°44'02,485" S	35°19'53,531" W
V-V-0141	V-V-0142	9.033.774.107	243.480.390	316°01'35"	54,738	1.00041465	8°44'02,038" S	35°19'53,486" W
V-V-0142	V-V-0143	9.033.813.500	243.442.384	242°53'41"	18,377	1.00041475	8°44'00,748" S	35°19'54,721" W
V-V-0143	V-V-0144	9.033.805.127	243.428.025	317°31'15"	2,961	1.00041476	8°44'01,018" S	35°19'55,258" W
V-V-0144	V-V-0145	9.033.807.310	243.424.028	344°22'28"	20,663	1.00041480	8°44'00,946" S	35°19'55,323" W
V-V-0145	V-V-0146	9.033.827.210	243.418.460	341°34'04"	32,822	1.00041486	8°44'00,298" S	35°19'55,501" W
V-V-0146	V-V-0147	9.033.858.348	243.408.082	343°22'35"	25,934	1.00041491	8°43'59,282" S	35°19'55,834" W
V-V-0147	V-V-0148	9.033.883.196	243.400.663	335°31'15"	30,180	1.00041499	8°43'58,472" S	35°19'56,071" W
V-V-0148	V-V-0149	9.033.910.647	243.388.166	349°59'37"	19,083	1.00041501	8°43'57,577" S	35°19'56,474" W
V-V-0149	V-V-0150	9.033.929.440	243.384.850	12°34'35"	7,394	1.00041500	8°43'56,965" S	35°19'56,579" W
V-V-0150	V-V-0151	9.033.936.656	243.386.460	15°37'31"	7,604	1.00041499	8°43'56,730" S	35°19'56,525" W
V-V-0151	V-V-0152	9.033.943.979	243.388.508	59°07'34"	17,657	1.00041489	8°43'56,492" S	35°19'56,456" W
V-V-0152	V-V-0153	9.033.953.040	243.403.663	76°18'08"	18,194	1.00041478	8°43'56,201" S	35°19'55,959" W
V-V-0153	V-V-0154	9.033.957.348	243.421.340	67°07'14"	17,031	1.00041468	8°43'56,064" S	35°19'55,380" W
V-V-0154	V-V-0155	9.033.963.970	243.437.031	48°09'18"	24,698	1.00041456	8°43'55,852" S	35°19'54,866" W
V-V-0155	V-V-0156	9.033.980.446	243.455.430	14°23'15"	35,939	1.00041451	8°43'55,319" S	35°19'54,261" W
V-V-0156	V-V-0157	9.034.015.258	243.464.360	322°09'06"	11,580	1.00041455	8°43'54,189" S	35°19'53,961" W
V-V-0157	V-V-0158	9.034.024.403	243.457.255	256°18'37"	11,557	1.00041462	8°43'53,890" S	35°19'54,192" W
V-V-0158	V-V-0159	9.034.021.668	243.446.026	201°30'01"	18,586	1.00041467	8°43'53,976" S	35°19'54,560" W
V-V-0159	V-V-0160	9.034.004.375	243.439.214	210°50'42"	14,872	1.00041472	8°43'54,538" S	35°19'54,786" W
V-V-0160	V-V-0161	9.033.991.606	243.431.569	235°10'53"	19,363	1.00041482	8°43'54,951" S	35°19'55,038" W
V-V-0161	V-V-0162	9.033.980.550	243.415.693	284°13'55"	35,163	1.00041503	8°43'55,308" S	35°19'55,560" W
V-V-0162	V-V-0163	9.033.969.195	243.381.610	308°04'21"	35,956	1.00041521	8°43'55,020" S	35°19'56,673" W
V-V-0163	V-V-0164	9.034.011.367	243.353.304	316°46'48"	27,017	1.00041533	8°43'54,293" S	35°19'57,594" W
V-V-0164	V-V-0165	9.034.031.055	243.334.803	306°57'38"	33,332	1.00041550	8°43'53,649" S	35°19'58,195" W
V-V-0165	V-V-0166	9.034.051.097	243.308.169	342°51'55"	26,367	1.00041555	8°43'52,991" S	35°19'59,062" W
V-V-0166	V-V-0167	9.034.076.294	243.300.401	10°37'08"	29,563	1.00041551	8°43'52,170" S	35°19'59,311" W
V-V-0167	V-V-0168	9.034.105.351	243.305.848	5°50'04"	33,462	1.00041549	8°43'51,225" S	35°19'59,127" W
V-V-0168	V-V-0169	9.034.138.639	243.309.250	13°47'47"	82,400	1.00041537	8°43'50,143" S	35°19'59,009" W
V-V-0169	V-V-0170	9.034.218.662	243.328.900	12°33'28"	55,689	1.00041529	8°43'47,543" S	35°19'58,350" W
V-V-0170	V-V-0171	9.034.273.018	243.341.008	359°34'43"	417,011	1.00041531	8°43'45,777" S	35°19'57,943" W
V-V-0171	V-V-0172	9.034.690.018	243.337.941	263°13'39"	85,621	1.00041584	8°43'32,209" S	35°19'57,959" W
V-V-0172	V-V-0173	9.034.709.610	243.254.592	278°17'16"	133,868	1.00041668	8°43'31,555" S	35°20'00,681" W
V-V-0173	V-V-0174	9.034.728.906	243.122.121	281°25'46"	226,066	1.00041810	8°43'30,901" S	35°20'05,009" W
V-V-0174	V-V-0175	9.034.774.104	242.896.558	304°49'53"	41,341	1.00041832	8°43'29,385" S	35°20'12,311" W
V-V-0175	V-V-0176	9.034.797.716	242.864.624	26°26'07"	6,240	1.00041830	8°43'28,610" S	35°20'13,415" W
V-V-0176	V-V-0177	9.034.803.303	242.867.402	25°41'48"	118,144	1.00041798	8°43'28,429" S	35°20'13,323" W
V-V-0177	V-V-0178	9.034.909.763	242.918.630	34°43'35"	102,663	1.00041760	8°43'24,975" S	35°20'11,627" W

```
V-V-0178 V-V-0179 9 034.994,140 242.977,113  33'41'30"  150,227  1 00041707  8'43'22,242" S  35'20'09,697" W
V-V-0179 V-V-0180 9 035.119,134 243.060,448  19'28'55"  170,309  1 00041671  8'43'18,192" S  35'20'06,947" W
V-V-0180 V-V-0181 9 035.279,693 243.117,247  18'13'37"  94,323   1 00041653  8'43'12,980" S  35'20'05,057" W
V-V-0181 V-V-0182 9 035.369,283 243.146,750  30'33'04"  296,894  1 00041557  8'43'10,071" S  35'20'04,074" W
V-V-0182 V-V-0001 9 035.824,961 243.297,663  22'32'51"  27,111   1 00041550  8'43'01,782" S  35'19'59,087" W
=====================================================================================================

                        Luciano André de Aguiar Pedrosa
Perimetro : 10 079,320 m             Engenheiro agrônomo
Área total  2 505 003,190 m²   250,5003 ha    CREA  PE 044 492 – R.N 180883771-1
=====================================================================================================
```

Source: UMBUZEIRO EMPREENDIMENTOS LTDA.

Based on this initial data, further analyses were carried out to define the property's panorama (Chart 3):

CAR - Rural Environmental Registry

IDENTIFICATION OF THE OWNER/POSSESSOR

CNPJ: 10.876.584/0001-17 - Name: UMBUZEIRO EMPREENDIMENTOS IMOBILIÁRIOS LTDA

Chart 3 - Overview of the property studied.

REAL ESTATE
Consolidated Area - 199.4099
Total property area - 250,5003
Remnant Native Vegetation - 50,4083
Administrative Servitude Area - 0,0000
Legal Reserve Area - 50,4083
Net Property Area - 250,5003
Permanent Preservation Area - 6.5771
Restricted Use Area - 0,0000

Source: Prepared by the author (2018).

DESCRIPTIVE MEMORIAL (UTM)

Property: ENGENHO BARRERINHO - PART 1

Owner: UMBUZEIRO EMPREENDIMENTOS IMOBILIÁRIOS LTDA

Municipality: TAMANDARÉ UF: PE - BR

Enrolment: IN PROGRESS

CNPJ: 10.876.584/0001-17

County: TAMANDARÉ

Area (ha): 250.5003

Perimeter (m): 10,079.320

The description of this perimeter begins at point V-V-0001, with coordinates N 9.035.650,004m and E 243.308,061m; from here it continues confronting ENGENHO SANTA CRUZ, with an azimuth of 152°45'24" for a distance of 37.163m to point V-

V-0002, with coordinates N 9.035.616.964m and E 243.325.073m; from this point continue with an azimuth of 152°45'24" for a distance of 155.575m to point V-V-0003, with coordinates N 9.035.478.646m and E 243.396.291m; from there continue with an azimuth of 152°45'24" for a distance of 569.853m to point V-V-0004, with coordinates N 9.034.972.007m and E 243.657.154m; from there continue with an azimuth of 163°27'16" for a distance of 193.103m to point V-V-0005, with coordinates N 9.034.786.900m and E 243.712.146m; from there continue with an azimuth of 146°53'51" for a distance of 122.467m to point V-V-0006, with coordinates N 9.034.684.310m and E 243.779.030m; from this point continue with an azimuth of 152°29'43" for a distance of 202.340m to point V-V-0007, with coordinates N 9.034.504.840m and E 243.872.475m; from this point continue with an azimuth of 151°30'05" for a distance of 229.630m to point V-V-0008, with coordinates N 9.034.303.035m and E 243.982.040m; from there continue with an azimuth of 151°30'05" for a distance of 108.289m to point V-V-0009, with coordinates N 9.034.207.868m and E 244.033.708m; from this point continue with an azimuth of 155°42'34" for a distance of 61.117m to point V-V-0010, with coordinates N 9.034.152.161m and E 244.058.850m; from this point continue with an azimuth of 145°37'39" for a distance of 50.467m to point V-V-0011, with coordinates N 9.034.110,506m and E 244.087,342m; from this point continue with an azimuth of 145°37'33" for a distance of 6.010m to point V-V-0012, with coordinates N 9.034.105,545m and E 244.090.735m; from there continue with an azimuth of 145°32'42" for a distance of 9.621m to point V-V-0013, with coordinates N 9.034.097.612m and E 244.096.179m; from there continue with an azimuth of 139°38'43" for a distance of 277.845m to point V-V-0014, with coordinates N 9.033.885.881m and E 244.276.088m; from there continue with an azimuth of 137°16'52" for a distance of 680.379m to point V-V-0015, with coordinates N 9.033.386.013m and E 244.737.658m; from there continue with an azimuth of 135°37'16" for a distance of 260.614m to point V-V-0016, with coordinates N 9.033.199.744m and E 244.919.932m; from there continue with an azimuth of 131°44'01" for a distance of 83.321m to point V-V-0017, with coordinates N

9.033.144.280m and E 244.982.110m; from there continue with an azimuth of 151°33'03" for a distance of 57.521m to point V-V-0018, with coordinates N 9.033.093.706m and E 245.009.511m; from there continue with an azimuth of 132°16'30" for a distance of 162.864m to point V-V-0019, with coordinates N 9.032.984.149m and E 245.130.018m; from there continue with an azimuth of 130°14'41" for a distance of 235.367m to point V-V-0020, with coordinates N 9.032.832.089m and E 245.309.673m; from there continue with an azimuth of 133°45'40" for a distance of 309.686m to point V-V-0021, with coordinates N 9.032.617.895m and E 245.533.337m; from there continue with an azimuth of 240°16'27" for a distance of 12.825m to point V-V-0022, with coordinates N 9.032.611.535m and E 245.522.200m; from there continue with an azimuth of 170°34'36" for a distance of 7.678m to point V-V-0023, with coordinates N 9.032.603.961m and E 245.523.457m; from there continue with an azimuth of 296°13'08" for a distance of 7.936m to vertex V-V-0024, with coordinates N 9.032.607.467m and E 245.516.338m; from there continue confronting with

ENGENHO VERMELHO 1, with an azimuth of 236°34'32" for a distance of 97.923m to point V-V-0025, with coordinates N 9.032.553,528m and E 245.434,610m; from this point continue with an azimuth of 235°49'40" for a distance of 76.767m to point V-V-0026, with coordinates N 9.032.510,409m and E 245.371,097m; from there continue with an azimuth of 242°11'48" for a distance of 46.585m to point V-V-0027, with coordinates N 9.032.488,680m and E 245.329,890m; then follow with an azimuth of 224°48'40" for a distance of 159.789m to point V-V-0028, with coordinates N 9.032.375,320m and E 245.217,275m; then follow with an azimuth of 195°18'04" for a distance of 52.397m to point V-V-0029, with coordinates N 9.032.324,780m and E 245.203.447m; from here continue with an azimuth of 201°39'38" for a distance of 36.486m to point V-V-0030, with coordinates N 9.032.290.870m and E 245.189.980m; from there continue with an azimuth of 195°54'45" for a distance of 170.649m to point V-V-0031, with coordinates N 9.032.126.760m and E 245.143.194m; from there continue with an azimuth of 150°16'49" for a distance of 25.850m to point V-V-0032, with coordinates N 9.032.104.310m and E 245.156.009m; from here it continues confronting ENGENHO SAUEZINHO, with an azimuth of 291°21'30" for a distance of 58.591m to vertex V-V- 0033, with coordinates N 9.032.125.649m and E 245.101.442m; from here it continues with an azimuth of 281°48'26" for a distance of 84.421m to vertex V-V- 0034, with coordinates N 9.032.142.923m and E 245.018.807m; from here continue with an azimuth of 290°20'59" for a distance of 27.210m to vertex V-V- 0035, with coordinates N 9.032.152.386m and E 244.993.294m; from here continue with an azimuth of 277°32'08" for a distance of 10.800m to vertex V-V- 0036, with coordinates N 9.032.153.802m and E 244.982.588m; from here continue with an azimuth of 271°24'40" for a distance of 37.256m to point V-V- 0037, with coordinates N 9.032.154.720m and E 244.945.342m; from here continue with an azimuth of 287°01'08" for a distance of 24.893m to point V-V- 0038, with coordinates N 9.032.162.005m and E 244.921.540m; from here continue with an azimuth of 297°08'36" for a distance of 12.828m to point V-V- 0039, with coordinates N 9.032.167.858m and E 244.910.124m; from there continue with an azimuth of 281°14'33" for a distance of 13.554m to point V-V- 0040, with coordinates N 9.032.170.500m and E 244.896.830m; from here continue with an azimuth of 288°31'56" for a distance of 15.032m to point V-V- 0041, with coordinates N 9.032.175.278m and E 244.882.577m; from here continue with an azimuth of 300°52'24" for a distance of 11.611m to point V-V- 0042, with coordinates N 9.032.181.236m and E 244.872.611m; from here continue with an azimuth of 289°48'03" for a distance of 45.418m to point V-V- 0043, with coordinates N 9.032.196.622m and E 244.829.879m; from here continue with an azimuth of 273°36'08" for a distance of 45.642m to point V-V- 0044, with coordinates N 9.032.199.490m and E 244.784.327m; from there continue with an azimuth of 283°40'57" for a distance of 14.629m to point V-V- 0045, with coordinates N 9.032.202.950m and E 244.770.113m; from here continue with an azimuth of 270°58'55" for a distance of 95.720m to point V-V- 0046, with coordinates N 9.032.204.590m and E 244.674.407m; from here continue with an azimuth of 246°08'51" for a distance of 15.165m to point V-V- 0047, with coordinates N 9.032.198.458m and E

244.660.538m; from here continue with an azimuth of 253°54'57" for a distance of 6.250m to vertex V-V- 0048, with coordinates N 9.032.196.727m and E 244.654.533m; from here continue with an azimuth of 320°00'47" for a distance of 9.512m to vertex V-V- 0049, with coordinates N 9.032.204.015m and E 244.648.420m; from there continue with an azimuth of 309°31'11" for a distance of 15.371m to point V-V- 0050, with coordinates N 9.032.213.796m and E 244.636.563m; from here continue with an azimuth of 295°55'39" for a distance of 18.781m to point V-V- 0051, with coordinates N 9.032.222.007m and E 244.619.672m; from there continue with an azimuth of 313°07'15" for a distance of 39.114m to point V-V- 0052, with coordinates N 9.032.248.743m and E 244.591.122m; from here continue with an azimuth of 294°22'29" for a distance of 9.036m to vertex V-V- 0053, with coordinates N 9.032.252.472m and E 244.582.892m; from here continue with an azimuth of 317°53'56" for a distance of 10.176m to vertex V-V- 0054, with coordinates N 9.032.260.023m and E 244.576.070m; from there continue with an azimuth of 312°03'50" for a distance of 34.510m to point V-V- 0055, with coordinates N 9.032.283.143m and E 244.550.449m; from this point continue with an azimuth of 279°36'31" for a distance of 36.404m to point V-V- 0056, with coordinates N 9.032.289.219m and E 244.514.556m; from this point continue with an azimuth of 298°23'42" for a distance of 14.980m to point V-V- 0057, with coordinates N 9.032.296.343m and E 244.501.378m; from here continue with an azimuth of 311°41'39" for a distance of 23.112m to vertex V-V- 0058, with coordinates N 9.032.311.716m and E 244.484.121m; from here continue with an azimuth of 317°44'26" for a distance of 6.250m to vertex V-V- 0059, with coordinates N 9.032.316,341m and E 244.479,918m; from there continue with an azimuth of 336°12'53" for a distance of 13.030m to point V-V- 0060, with coordinates N 9.032.328.265m and E 244.474.662m; from this point continue with an azimuth of 360°00'00" for a distance of 12.579m to point V-V- 0061, with coordinates N 9.032.340.844m and E 244.474.662m; from there continue with an azimuth of 331°19'38" for a distance of 29.459m to point V-V- 0062, with coordinates N 9.032.366.690m and E

244.460.528m; from here continue with an azimuth of 339°58'11" for a distance of 9.345m to point V-V- 0063, with coordinates N 9.032.375.470m and E 244.457.327m; from here continue with an azimuth of 324°31'44" for a distance of 21.793m to point V-V- 0064, with coordinates N 9.032.393.218m and E 244.444.681m; from there continue with an azimuth of 340°57'04" for a distance of 10.999m to point V-V- 0065, with coordinates N 9.032.403.615m and E 244.441.091m; from this point continue with an azimuth of 353°30'33" for a distance of 9.914m to point V-V- 0066, with coordinates N 9.032.413.466m and E 244.439.970m; from here continue with an azimuth of 8°00'26" for a distance of 7.968m to point V-V- 0067, with coordinates N 9.032.421.357m and E 244.441.080m; from here continue with an azimuth of 356°03'53" for a distance of 13.652m to point V-V- 0068, with coordinates N 9.032,434.977m and E 244,440.143m; from there continue with an azimuth of 347°22'39" for a distance of 11.337m to point V-V- 0069, with coordinates N 9.032.446.040m and E 244.437.666m; from there continue with an azimuth of 334°26'47" for a distance of 9.170m to point V-V- 0070, with coordinates N 9.032.454.313m and E 244.433.710m; from this point continue with an azimuth of 8°18'11" for a distance of 11.495m to point V-V- 0071, with coordinates N 9.032.465.688m and E 244.435.370m; from here continue with an azimuth of 335°05'37" for a distance of 2.217m to point V-V- 0072, with coordinates N 9.032.467.699m and E 244.434.436m; from here continue with an azimuth of 360°00'00" for a distance of 2.755m to point V-V- 0073, with coordinates N 9.032.470.454m and E 244.434.436m; from there continue with an azimuth of 8°24'34" for a distance of 16.231m to point V-V- 0074, with coordinates N 9.032.486.510m and E 244.436.810m; from this point continue with an azimuth of 16°13'50" for a distance of 8.515m to point V-V- 0075, with coordinates N 9.032.494.686m and E 244.439.190m; from here continue with an azimuth of 348°29'13" for a distance of 8.066m to point V-V- 0076, with coordinates N 9.032.502.590m and E 244.437.580m; from here continue with an azimuth of 342°04'24" for a distance of 13.905m to point V-V- 0077, with coordinates N 9.032.515.820m and E 244.433.300m; from here continue with an azimuth of 327°04'33" for a distance of 9.475m to point V-V- 0078, with coordinates N 9.032.523.773m and E 244.428.150m; from there continue with an azimuth of 345°27'28" for a distance of 11.474m to point V-V- 0079, with coordinates N 9.032.534.880m and E 244.425.269m; from there continue with an azimuth of 333°02'32" for a distance of 25.387m to point V-V- 0080, with coordinates N 9.032.557.509m and E 244.413.760m; from this point continue with an azimuth of 304°11'29" for a distance of 28.652m to point V-V- 0081, with coordinates N 9.032.573.610m and E 244.390.060m; from here continue with an azimuth of 353°47'00" for a distance of 35.834m to point V-V- 0082, with coordinates N 9.032.609.233m and E 244.386.180m; from here continue with an azimuth of 274°26'05" for a distance of 7.239m to point V-V- 0083, with coordinates N 9.032.609.793m and E 244.378.962m; from this point continue with an azimuth of 323°37'54" for a distance of 9.540m to point V-V- 0084, with coordinates N 9.032.617.475m and E 244.373.306m; from this point continue with an azimuth of 304°32'26" for a distance of 6.412m to point V-V- 0085, with coordinates N 9.032.621.110m and E 244.368.024m; from there continue with an azimuth of 322°40'43" for a distance of 24.533m to point V-V- 0086, with coordinates N 9.032.640.620m and E 244.353.150m; from here continue with an azimuth of 317°05'25" for a distance of 19.292m to point V-V- 0087, with coordinates N 9.032.654.750m and E 244.340.015m; from here continue with an azimuth of 354°59'48" for a distance of 18.862m to point V-V- 0088, with coordinates N 9.032.673.540m and E 244.338.370m; from here continue with an azimuth of 318°55'02" for a distance of 20.619m to point V-V- 0089, with coordinates N 9.032.689.082m and E 244.324.820m; from this point continue with an azimuth of 328°21'09" for a distance of 14.740m to point V-V- 0090, with coordinates N 9.032.701.630m and E 244.317.086m; from this point continue with an azimuth of 6°28'57" for a distance of 21.820m to point V-V- 0091, with coordinates N 9.032.723.310m and E 244.319.550m; from here continue with an azimuth of 340°50'27" for a distance of 15.647m to point V-V- 0092, with coordinates N 9.032.738.090m and E 244.314.415m; from here continue with an azimuth of 330°25'56" for a distance of 24.409m to point V-V- 0093, with coordinates N

9.032.759.320m and E 244.302.370m; from here continue with an azimuth of 335°06'31" for a distance of 13.846m to point V-V- 0094, with coordinates N 9.032.771.880m and E 244.296.542m; from there continue with an azimuth of 342°29'05" for a distance of 10.181m to point V-V- 0095, with coordinates N 9.032.781.589m and E 244.293.478m; from here continue with an azimuth of 335°55'16" for a distance of 9.750m to point V-V- 0096, with coordinates N 9.032.790.491m and E 244.289.500m; from here continue with an azimuth of 317°13'33" for a distance of 9.699m to point V-V- 0097, with coordinates N 9.032.797.610m and E 244.282.914m; from here continue with an azimuth of 338°33'03" for a distance of 22.552m to point V-V- 0098, with coordinates N 9.032.818.599m and E 244.274.667m; from here continue with an azimuth of 314°39'20" for a distance of 9.302m to point V-V- 0099, with coordinates N 9.032.825.137m and E 244.268.050m; from there continue with an azimuth of 335°59'54" for a distance of 9.253m to point V-V- 0100, with coordinates N 9.032.833.590m and E 244.264.286m; from this point continue with an azimuth of 350°59'56" for a distance of 14.207m to point V-V- 0101, with coordinates N 9.032.847.622m and E 244.262.064m; from here continue with an azimuth of 329°47'34" for a distance of 37.409m to point V-V- 0102, with coordinates N 9.032.879.952m and E 244.243.242m; from here continue with an azimuth of 316°00'05" for a distance of 33.517m to point V-V- 0103, with coordinates N 9.032.904.062m and E 244.219.960m; from here continue with an azimuth of 301°36'21" for a distance of 40.860m to point V-V- 0104, with coordinates N 9.032.925.476m and E 244.185.161m; from there continue with an azimuth of 301°31'48" for a distance of 11.002m to point V-V- 0105, with coordinates N 9.032.931.229m and E 244.175.783m; from this point continue with an azimuth of 325°31'03" for a distance of 8.724m to point V-V- 0106, with coordinates N 9.032.938.420m and E 244.170.844m; from this point continue with an azimuth of 303°26'06" for a distance of 132.634m to point V- V-0107, with coordinates N 9.033.011.500m and E 244.060.160m; from here continue with an azimuth of 295°27'20" for a distance of 47.253m to point V-V- 0108, with coordinates N

9.033.031.810m and E 244.017.494m; from here continue with an azimuth of 266°01'59" for a distance of 16.489m to point V-V- 0109, with coordinates N 9.033.030.669m and E 244.001.044m; from there continue with an azimuth of 309°52'25" for a distance of 112.467m to point V- V-0110, with coordinates N 9.033.102.772m and E 243.914.730m; from here continue with an azimuth of 298°19'03" for a distance of 63.367m to point V-V- 0111, with coordinates N 9.033.132.830m and E 243.858.946m; from here it continues confronting ENGENHO BARRERINHO PARTE 2 (SILVIO ROMERO), with an azimuth of 37°47'47" for a distance of 59.531m to vertex V-V- 0112, with coordinates N 9.033.179.871m and E 243.895.430m; from this point continue with an azimuth of 22°28'47" for a distance of 27.252m to point V-V- 0113, with coordinates N 9.033.205.052m and E 243.905.850m; from there continue with an azimuth of 341°23'44" for a distance of 21.252m to point V-V- 0114, with coordinates N 9.033.225.193m and E 243.899.070m; from here continue with an azimuth of 349°38'38" for a distance of 17.356m to point V-V- 0115, with coordinates N 9.033.242.266m and E 243.895.950m; from here continue with an azimuth of 358°42'38" for a distance of 21.777m to point V-V- 0116, with coordinates N 9.033.264.038m and E 243.895.460m; from there continue with an azimuth of 5°39'44" for a distance of 13.257m to point V-V- 0117, with coordinates N 9.033.277.230m and E 243.896.768m; from this point continue with an azimuth of 17°27'53" for a distance of 11.306m to point V-V- 0118, with coordinates N 9.033.288.015m and E 243.900.161m; from this point continue with an azimuth of 13°16'24" for a distance of 40.892m to point V-V- 0119, with coordinates N 9.033.327.815m and E 243.909.550m; from here continue with an azimuth of 6°40'20" for a distance of 31.972m to point V-V- 0120, with coordinates N 9.033.359.570m and E 243.913.265m; from here continue with an azimuth of 338°46'39" for a distance of 87.150m to point V-V- 0121, with coordinates N 9.033.440.810m and E 243.881.717m; from there continue with an azimuth of 346°09'42" for a distance of 27.414m to point V-V- 0122, with coordinates N 9.033.467.429m and E 243.875.160m; from there continue with an azimuth of 356°25'52" for a distance of

72.452m to point V-V- 0123, with coordinates N 9.033.539.740m and E 243.870.650m; from here continue with an azimuth of 351°24'21" for a distance of 66.049m to point V-V- 0124, with coordinates N 9.033.605.048m and E 243.860.780m; from here continue with an azimuth of 358°26'04" for a distance of 10.615m to point V-V- 0125, with coordinates N 9.033.615.659m and E 243.860.490m; from here continue with an azimuth of 0°17'33" for a distance of 17.627m to point V-V- 0126, with coordinates N 9.033.633.286m and E 243.860.580m; from there continue with an azimuth of 13°53'47" for a distance of 61.790m to point V-V- 0127, with coordinates N 9.033.693.268m and E 243.875.420m; from this point continue with an azimuth of 15°08'13" for a distance of 32.153m to point V-V- 0128, with coordinates N 9.033.724.306m and E 243.883.816m; from here continue with an azimuth of 247°46'42" for a distance of 13.558m to point V-V- 0129, with coordinates N 9.033.719.178m and E 243.871.265m; from here continue with an azimuth of 260°37'24" for a distance of 10.650m to point V-V- 0130, with coordinates N 9.033.717.443m and E 243.860.757m; from here continue with an azimuth of 285°22'15" for a distance of 77.102m to point V-V- 0131, with coordinates N 9.033.737.880m and E 243.786.413m; from there continue with an azimuth of 278°07'23" for a distance of 60.267m to point V-V- 0132, with coordinates N 9.033.746.396m and E 243.726.751m; from here continue with an azimuth of 279°06'33" for a distance of 56.118m to point V-V- 0133, with coordinates N 9.033.755.280m and E 243.671.340m; from here continue with an azimuth of 252°14'05" for a distance of 55.547m to point V-V- 0134, with coordinates N 9.033.738.332m and E 243.618.442m; from here continue with an azimuth of 249°59'07" for a distance of 11.838m to point V-V- 0135, with coordinates N 9.033.734.280m and E 243.607.319m; from here continue with an azimuth of 258°08'06" for a distance of 63.569m to point V-V- 0136, with coordinates N 9.033.721.210m and E 243.545.109m; from there continue with an azimuth of 254°47'25" for a distance of 29.388m to point V-V- 0137, with coordinates N 9.033.713.500m and E 243.516.750m; from here continue with an azimuth of 276°23'52" for a distance of

28.478m to point V-V- 0138, with coordinates N 9.033.716.673m and E 243.488.450m; from here continue with an azimuth of 358°37'17" for a distance of 27.585m to point V-V- 0139, with coordinates N 9.033.744.250m and E 243.487.786m; from here continue with an azimuth of 331°38'47" for a distance of 18.312m to point V-V- 0140, with coordinates N 9.033.760.365m and E 243.479.090m; from here continue with an azimuth of 5°24'14" for a distance of 13.804m to point V-V- 0141, with coordinates N 9.033.774.107m and E 243.480.390m; from there continue with an azimuth of 316°01'35" for a distance of 54.738m to point V-V- 0142, with coordinates N 9.033.813.500m and E 243.442.384m; from this point continue with an azimuth of 242°53'41" for a distance of 18.377m to point V-V- 0143, with coordinates N 9.033.805.127m and E 243.426.025m; from there continue with an azimuth of 317°31'15" for a distance of 2.961m to point V-V- 0144, with coordinates N 9.033.807.310m and E 243.424.026m; from here continue with an azimuth of 344°22'28" for a distance of 20.663m to point V-V- 0145, with coordinates N 9.033.827.210m and E 243.418.460m; from here continue with an azimuth of 341°34'04" for a distance of 32.822m to point V-V- 0146, with coordinates N 9.033.858.348m and E 243.408.082m; from there continue with an azimuth of 343°22'35" for a distance of 25.934m to point V-V- 0147, with coordinates N 9.033.883.198m and E 243.400.663m; from this point continue with an azimuth of 335°31'15" for a distance of 30.160m to point V-V- 0148, with coordinates N 9.033.910.647m and E 243.388.166m; from here continue with an azimuth of 349°59'37" for a distance of 19.083m to point V-V- 0149, with coordinates N 9.033.929.440m and E 243.384.850m; from there continue with an azimuth of 12°34'35" for a distance of 7.394m to point V-V- 0150, with coordinates N 9.033.936.656m and E 243.386.460m; from there continue with an azimuth of 15°37'31" for a distance of 7.604m to point V-V- 0151, with coordinates N 9.033.943.979m and E 243.388.508m; from there continue with an azimuth of 59°07'34" for a distance of 17.657m to point V-V- 0152, with coordinates N 9.033.953.040m and E 243.403.663m; from this point continue with an azimuth of 76°18'08" for a distance of 18.194m to point V-V- 0153, with coordinates N

9.033.957.348m and E 243.421.340m; from here continue with an azimuth of 67°07'14" for a distance of 17.031m to point V-V- 0154, with coordinates N 9.033.963.970m and E 243.437.031m; from here continue with an azimuth of 48°09'18" for a distance of 24.698m to point V-V- 0155, with coordinates N 9.033.980.446m and E 243.455.430m; from here continue with an azimuth of 14°23'15" for a distance of 35.939m to point V-V- 0156, with coordinates N 9.034.015.258m and E 243.464.360m; from there continue with an azimuth of 322°09'06" for a distance of 11.580m to point V-V- 0157, with coordinates N 9.034.024.403m and E 243.457.255m; from this point continue with an azimuth of 256°18'37" for a distance of 11.557m to point V-V- 0158, with coordinates N 9.034.021.668m and E 243.446.026m; from here continue with an azimuth of 201°30'01" for a distance of 18.586m to point V-V- 0159, with coordinates N 9.034.004.375m and E 243.439.214m; from here continue with an azimuth of 210°50'42" for a distance of 14.872m to point V-V- 0160, with coordinates N 9.033.991.606m and E 243.431.589m; from here continue with an azimuth of 235°10'53" for a distance of 19.363m to point V-V- 0161, with coordinates N 9.033.980.550m and E 243.415.693m; from there continue with an azimuth of 284°13'55" for a distance of 35.163m to point V-V- 0162, with coordinates N 9.033.989.195m and E 243.381.610m; from there continue with an azimuth of 308°04'21" for a distance of 35.956m to point V-V- 0163, with coordinates N 9.034.011.367m and E 243.353.304m; from here continue with an azimuth of 316°46'48" for a distance of 27.017m to point V-V- 0164, with coordinates N 9.034.031.055m and E 243.334.803m; from here continue with an azimuth of 306°57'38" for a distance of 33.332m to point V-V- 0165, with coordinates N 9.034.051.097m and E 243.308.169m; from here continue with an azimuth of 342°51'55" for a distance of 26.367m to point V-V- 0166, with coordinates N 9.034.076,294m and E 243.300,401m; then follow with an azimuth of 10°37'08" for a distance of 29.563m to point V-V- 0167, with coordinates N 9.034.105,351m and E 243.305,848m; then follow with an azimuth of 5°50'04" for a distance of 33.462m to point V-V- 0168, with coordinates N 9.034.138,639m and E 243.309.250m; from here continue with an azimuth of 13°47'47" for a distance of 82.400m to point V-V- 0169, with coordinates N 9.034.218.662m and E 243.328.900m; from here continue with an azimuth of 12°33'28" for a distance of 55.689m to point V-V- 0170, with coordinates N 9.034.273.018m and E 243.341.008m; from there continue with an azimuth of 359°34'43" for a distance of 417.011m to point V- V-0171, with coordinates N 9.034.690.018m and E 243.337.941m; from there continue with an azimuth of 283°13'39" for a distance of 85.621m to point V-V- 0172, with coordinates N 9.034.709.610m and E 243.254.592m; from there continue with an azimuth of 278°17'16" for a distance of 133.868m to point V- V-0173, with coordinates N 9.034.728.906m and E 243.122.121m; from here continue with an azimuth of 281°25'46" for a distance of 228.086m to point V- V-0174, with coordinates N 9.034.774.104m and E 242.898.558m; from here continue with an azimuth of 304°49'53" for a distance of 41.341m to point V-V- 0175, with coordinates N 9.034.797.716m and E 242.864.624m; from here continue with an azimuth of 26°26'07" for a distance of 6.240m to point V-V- 0176, with coordinates N 9.034.803,303m and E 242.867,402m; from this point, confronting ENHENHO JOSÉ DA COSTA, with an azimuth of 25°41'48" for a distance of 118.144m to point V-V-0177, with coordinates N 9.034.909.763m and E 242.918.630m; from there continue with an azimuth of 34°43'35" for a distance of 102.663m to point V-V-0178, with coordinates N 9.034.994,140m and E 242.977,113m; from there continue with an azimuth of 33°41'30" for a distance of 150.227m to point V-V-0179, with coordinates N 9.035.119,134m and E 243.060.448m; from there continue with an azimuth of 19°28'55" for a distance of 170.309m to point V-V-0180, with coordinates N 9.035.279.693m and E 243.117.247m; from there continue with an azimuth of 18°13'37" for a distance of 94.323m to point V-V-0181, with coordinates N 9.035.369.283m and E 243.146.750m; from here continue with an azimuth of 30°33'04" for a distance of 296.894m to point V-V-0182, with coordinates N 9.035.624.961m and E 243.297.663m; from here continue with an azimuth of 22°32'51" for a distance of 27.111m to point V-V-0001, the starting point for the description of this perimeter.

All the coordinates described here are georeferenced to the Brazilian Geodetic System and are represented in the UTM System, referenced to Central Meridian No. 33 WGr, with SIRGAS2000 as the Datum. All azimuths and distances, area and perimeter were calculated in the UTM projection plane. The plan of Engenho Barreirinho was then drawn up in TOPODATA (Figure 4):

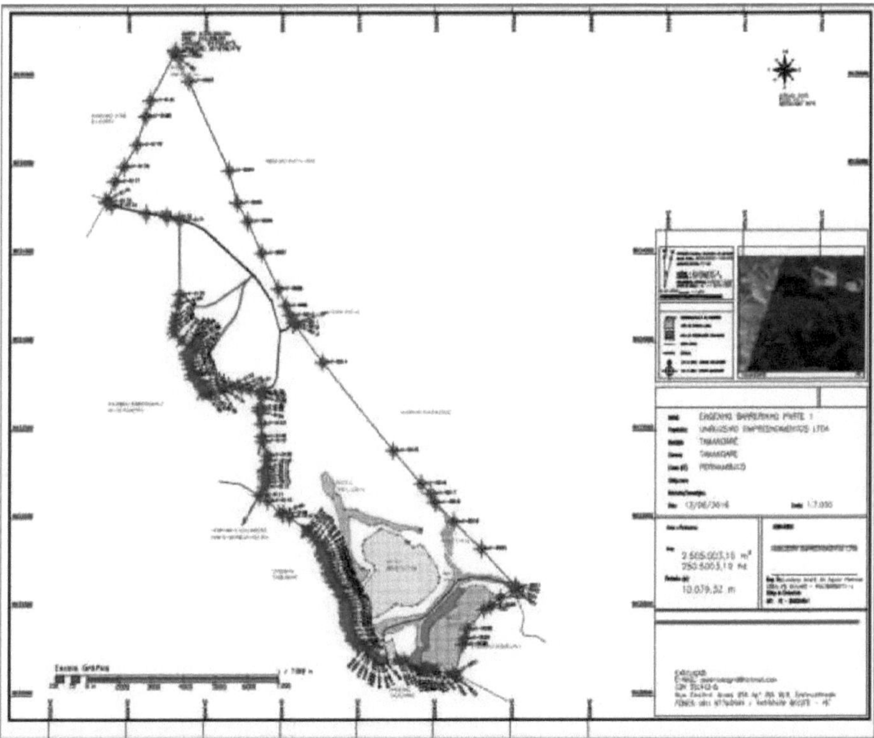

Figure 4 - Plan of the Barreirinho sugar mill.
Source: Prepared by the author (2018).

CHAPTER 6

FINAL CONSIDERATIONS

Based on the SAVI, the work showed that most of the municipality of Tamandaré-PE is anthropised and has sparse vegetation (characteristic of agricultural crops in the area). In this context, the property was geo-referenced and then the CAR was drawn up in which the type of land use on the property was surveyed. Subsequently, a descriptive memorial was drawn up and a plan of the rural property was drawn up.

The work therefore proved useful for analysing the property in relation to the region it comprises. These results have the potential to serve as a product for georeferencing and analysing the rural property belonging to Engenho Barreirinho.

REFERENCES

ALLEN, R.; TASUMI, M.; TREZZA, R. SEBAL (Surface Energy Balance Algorithms for Land). **Advanced Training and User's Manual -** Idaho Implementation, version 1.0. 2002.

AVERY, T. E.; BERLIN, G. L. **Fundamentals of remote sensing and airphoto interpretation.** 5. ed. New Jersey: Prentice Hall, 1992.

ARAÚJO, M. C. B., & COSTA, M. F. **Quali-quantitative analysis of the rubbish left in Tamandaré Bay - PE - Brazil by excursionists.** Integrated Coastal Management, 3 (1): 58-61. 2003.

BORGES, M.H. PFEIFER R.M. DEMATTÊ J.A.M. Evolução e mapeamento do uso da terra, através de imagens aerofotogramétricas e orbitais em Santa Bárbara D'Oeste (SP). **Scientia agrícola,** 50 (3): 365-371 Oct./Dec., 1993.

BRANDALIZE, M.C. B.. **Geoprocessing: Notes.** CURITIBA, Federal University of Paraná, 2008.

BRINKER, R. C.; WOLF, P. R. **Elementary Surveying.** 6 ed. New York: Harper &

Row, 568 p. 1977.

CAMPOS, SÉRGIO; ARAÚJO JÚNIOR, A. A.; BARROS, Z. X.; CARDOSO. L. G; PIROLI, E.L. Remote sensing and geoprocessing applied to land use in micro-watersheds, Botucatu-SP. **Eng. Agríc.**, Jaboticabal, v.24, n.2, p.431-435, May/Aug. 2004.

CARVALHO JÚNIOR OA, GUIMARÃES RF, CARVALHO APF, GOMES RAT, MELO AF, SILVA PA. **Processing and analysing multitemporal images for the Gorutuba/MG irrigation perimeter.** In: Actas XII SIMPÓSIO BRASILEIRO DE SENSORIAMENTO REMOTO [internet]; 2005 April 16-21; Goiânia, Brazil. 2005 [access in 2017 January 13].

Available at   :
http://marte.sid.inpe.br/col/ltid.inpe.br/sbsr/2004/12.06.13.32/doc/473.pdf. Accessed: 15/8/2017.

CASTRO, T. S.; SILVA, M. V. da. **Using remote sensing data to delimit rural properties according to the technical standard for georeferencing rural properties.** 3ª edition. 2014.

COELHO, A. G. S. **Aerial Photographs in the Classification of Agricultural Land. São Paulo.** Institute of Geography. USP. Aerophotography Magazine. No. 6, 13 p. 1971.

CONCEIÇÃO, L. A. B. da SILVA. **Environmental diagnosis through the use of remote sensing techniques as support for the planning of administrative units: the case of Osório, RS.** Dissertation. Federal University of Rio Grande do Sul, 2004.

CUIABANO, J. L. S. P. **Basic Technical Drawing Handouts.** Class notes. Cuiabá : Federal University of Mato Grosso, 2006.

DAIANESE, R. C. **Remote sensing and geoprocessing applied to the temporal study of land use and the comparison between non-supervised classification and visual analysis.** 2001. 186 f. Dissertation (Master's Degree in Agronomy/Energia na Agricultura) -

Faculty of Agronomic Sciences, São Paulo State University, Botucatu, 2001.

DIEGUES, A.C. **Sustainable development, geo-environmental and natural resource management.** CADERNOS FUNDAP, São Paulo, Year 9 No. 16 - p. 33-45 . jun. /1989.

DOUBECK, A. **Topography.** Curitiba: Federal University of Paraná, 1989.

ESPARTEL, L. **Curso de Topografia.** 9 ed. Rio de Janeiro: Globo, 1987.

FACCO, D. S.; BENEDETTI, A. C.P.; PEREIRA FILHO, W.; KAISER, E.A.; DAL OSTO, J. V. Geotechnologies for forest monitoring in the municipality of Nova Palma - Rio Grande Do Sul - BR. **Revista Eletrónica em Gestão, Educação e Tecnologia Ambiental Santa Maria, v.** 20, n. 1, p. 417-426. DOI: 105902/22361170 19946. Journal of the Centre for Natural and Exact Sciences - UFSM ISSN : 22361170. Jan.-Apr. 2016.

FAGGION, P. L; VEIGA, L. A. K; ZANETTI, M. A. **Fundamentals of Topography.** Paraná: Federal University of Paraná, 2012.

FUCHS, R. B. H. **Evaluation of land use, by slope class, in the Vacacaí-Mirtim River sub-basin / RS.** Santa Maria /RS. 59 p. Specialisation Monograph. UFSM. 1986.

GARCIA, G.J.; PIEDADE, G.C.R. **Topography Applied to Agricultural Sciences.** 5ª ed.

GUERREIRO, J. **What is GPS?** Typed. 18p. 2002.

HUETE, A. R. Adjusting vegetation indices for soil influences. International Agrophysics, v.4, n.4, p.367-376, 1988.

IBGE. **Population of Tamandaré by race and colour.** IBGE Automatic Retrieval System (SIDRA), (2010). Consulted on 20 January 2017.

IBGE. **Official territorial area.** Brazilian Institute of Geography and Statistics, 2013. Consulted on 20 January 2017.

INCRA. NATIONAL INSTITUTE FOR COLONISATION AND AGRARIAN REFORM. **Technical standard for georeferencing rural properties.** 3rd Edition. Brasília, 2013. Available at: https://sigef.incra.gov.br/static/documentos/norma_tecnica_georreferenciame nto_imoveis_rurais_3ed.pdf

INCRA. NATIONAL INSTITUTE FOR COLONISATION AND AGRARIAN REFORM. **Technical Manual on Boundaries and Confrontations.** Brasília, 1ª Edition. 2013.

INPE. National Institute for Space Research. **Monitoring of the Brazilian Amazonian forest by satellite,** 2000- 2001. São José dos Campos, SP. 2002.

IPPOLITI-RAMILO G. A, EPIPHANIO J. C. N & SHIMABUKURO Y. E. Landsat-5 Thematic Mapper data for pre-planting crop area evaluation in tropical countries. **International Journal of Remote Sensing** 24:1521-1534. 2003.

JENSEN, J. R. **Remote sensing of the environment:** an Earth Resources Perspective. 2. ed. Upper Saddle River: PrenticeHall. 592p. 2007.

LIMA, S. H. **Age, growth and some aspects of the reproduction of Stegastes fuscus, Cuvier 1830 (Teleostei, Pomacentridae) from the reefs of Tamandaré, Pernambuco. Brazil.** Master's dissertation. 1997.

MARKHAM, B.L.; BARKER, L.L. Thematic mapper bandpass solar exoatmospherical irradiances. **International Journal of Remote Sensing,** v.8, n.3, p.517-523. 1987.

MENESES, P. R. **Fundamentals of Optical Spectral Radiometry.** In: MENESES, P. R.; NETTO, J. S. M. **Sensoriamento Remoto:** Reflectância dos alvos naturais. Brasília, DF: UnB; Planaltina: Embrapa Cerrados. 2001.

MENESES, P. R; ALMEIDA, T. D. **Introduction to Digital Remote Sensing Image Processing.** BRASÍLIA. 2010.

NOVO, E. M. **Remote sensing: principles and applications**. São Paulo. Edgard Blucher, 309p. 1989.

OLIVEIRA, C. L. The **Fair Distribution of Land.** Available at < www.geodesia.ufsc.br/wiki-ctm/index.php/A_distribuição_justa_da_terra >. Accessed on 01/01/2017

OLIVEIRA, T. H.; SILVA, J. S.; MACHADO C. C. C.; GALVÍNCIO, J. D.; PIMENTEL, R. M. M.; SILVA, B. B. Moisture index (NDWI) and spatio-temporal analysis of the surface albedo of the Moxotó river basin in Pernambuco. **Brazilian Journal of Physical Geography.** V.03, p.55 - 69, Homepage: www.ufpe.br/rbgfe. 2010.

PAULA, M. R. de; CABRAL, J. B. P.; MARTINS, A. P. Use of remote sensing and geoprocessing techniques to characterise land use in the Caçu HPP Hydrographic Basin - GO. **REVISTA GEONORTE**, Special Issue, V.4, N.4, p.1482 - 1490, 2012.

PEREIRA, M. N., KURKDJIAN, M. L. N. O. ; FORESTI, C. **Land cover and land use through remote sensing.** São José dos Campos, Space Research Institute. 118 p. 1989.

RIZZI R. & RUDORFF B.F.T. Estimation of soya area in Rio Grande do Sul using Landsat images. **Revista Brasileira de Cartografia** 57:226-234. 2005.

ROCHA, J. S. M. **Contribution to the quantitative evaluation of land use capacity in the State of Rio Grande do Sul.** Santa Maria, 169 p. Honours thesis. Federal University of Santa Maria. 1977.

ROQUE, C.G.; OLIVEIRA, I. C. de; PRISCILA PEREIRA FIGUEIREDO , EVERTON VALDOMIRO PEDROSO BRUM E MAIRO FABIO CAMARGO.

Georeferencing. **Revista de Ciências Agro-Ambientais,** Alta Floresta, v.4, n.1, p.87-102, 2006.

ROSA, O. **Land use map of the municipality of Santa Maria - RS,** adapted for children in the 4th grade. Santa Maria. 72 p. (monograph) 1994.

SILVA, E. G. da. Area measurements from aerial photographs, on a nominal scale, compared with the area obtained from photos with corrected scales using a GIS. **Energy in Agriculture.** V.26, n. 2 (2009). Botucatu - SP November - 2009.

SILVA, E. R. A. C.; GALVÍNCIO, J. D. The MWSP Global Scope Methodology Applied at the Local Level to Analyse Hydrological Stress in the Middle Stretch of the Ipojuca Basin - PE: a Contribution to the Theme of the Transposition of the São Francisco River. **Revista Brasileira de Geografia Física,** Vol 4 n° 03. 602-628. 2011a.

SILVA, E. R. A. C.; MELO, J. G.; GALVINCIO, J. D. Identification of Areas Susceptible to Desertification Processes in the Middle Section of the Ipojuca Basin - PE through the Mapping of Vegetation Water Stress and the Estimation of the Aridity Index. **Revista Brasileira de Geografia Física,** Vol. 4, No 3, p. 629-649. 2011b.

SILVA, E. R. A. C.; MORAIS, Y. C. B. ; SILVA, J. F. ; GALVINCIO, J. D. . Water consumption irrigation for banana farming in edaphoclimatic conditions of the stream of Pontal basin in Semiarid of Pernambuco. **Revista Brasileira de Geografia Física**, v. 8, p. 921-937, 2015.

SILVA, E.R.A.C., GALVÍNCIO, J.D., BRANDÃO NETO, J.L., MORAIS, Y.C.B. Space-Time Analysis of Environmental Changes and their Reflection on the Development of Phenological of Vegetation of Mangrove. Journal of

Agriculture and Environmental Sciences 4,   245-253. DOI: 10.15640/jaes.v4n1a30. 2015.

SILVA, E.R.A.C.; MIRANDA, R. Q.; FERREIRA, P. dos S.; GOMES, V. P. GALVÍNCIO, J.D. Estimation of Hydrological Stress in the Riacho do Pontal-PE Watershed. ISSN 2318-2962. DOI 10.5752/p.2318- 2962.2016v26n47p844. **Caderno de Geografia,** v.26, n.47, 2016.

STEINER, D. Time dimension for crop surveys from space. Photogrammetric Engineering. Falls Church, v.36, n.2, p.187-194. 1970.

TUCCI, C.E.M. **Hidrologia:** ciência e aplicação. Porto Alegre: Editora da Universidade/ABRH, ch.1, p.25-33; ch.22, p.849-75. 1993.

VALENTE, O.F.; CASTRO, P.S. A bacia hidrográfica e a produção de água. **Informe Agropecuário**, Belo Horizonte, v.9, p.54-6, 1983.

VEIGA, L. A. K. ZANETTI, M. A. Z.; FAGGION, P. L. **Fundamentals of Topography.** Curitiba: Federal University of Paraná, 2007.

VIGANÓ, H. A; BORGES, E. F. FRANCA-ROCHA, W. J. S. **Analysing the performance of NDVI and SAVI vegetation indices from Aster images.** [electronic version] Proceedings... XV Brazilian Symposium on Remote Sensing - SBSR, Curitiba, PR, Brazil, INPE p.1828. 2011.

VOGEL, E.; MARQUES, F. P.; ROCHA, I. R.; OLIVEIRA, R. C.; SARAIVA, C. C. S. **Case study of an altimetric topographic survey carried out with a total station and terrestrial laser scanning.** Available at: http://mundogeo.com/blog/2011/09/08/estudo-de-caso-de-um-levantamento-topografico-altimetrico-realizado-com-estacao-total-e-laser-scanning- terrestre/ Accessed on 11/01/2017

Printed by Books on Demand GmbH, Norderstedt / Germany